SpringerBriefs in Applied Sciences and Technology

Manufacturing and Surface Engineering

Series Editor

Joao Paulo Davim ⓘ, Department of Mechanical Engineering, University of Aveiro, Aveiro, Portugal

This series fosters information exchange and discussion on all aspects of manufacturing and surface engineering for modern industry. This series focuses on manufacturing with emphasis in machining and forming technologies, including traditional machining (turning, milling, drilling, etc.), non-traditional machining (EDM, USM, LAM, etc.), abrasive machining, hard part machining, high speed machining, high efficiency machining, micromachining, internet-based machining, metal casting, joining, powder metallurgy, extrusion, forging, rolling, drawing, sheet metal forming, microforming, hydroforming, thermoforming, incremental forming, plastics/composites processing, ceramic processing, hybrid processes (thermal, plasma, chemical and electrical energy assisted methods), etc. The manufacturability of all materials will be considered, including metals, polymers, ceramics, composites, biomaterials, nanomaterials, etc. The series covers the full range of surface engineering aspects such as surface metrology, surface integrity, contact mechanics, friction and wear, lubrication and lubricants, coatings an surface treatments, multi-scale tribology including biomedical systems and manufacturing processes. Moreover, the series covers the computational methods and optimization techniques applied in manufacturing and surface engineering. Contributions to this book series are welcome on all subjects of manufacturing and surface engineering. Especially welcome are books that pioneer new research directions, raise new questions and new possibilities, or examine old problems from a new angle. To submit a proposal or request further information, please contact Dr. Mayra Castro, Publishing Editor Applied Sciences, via mayra.castro@springer.com or Professor J. Paulo Davim, Book Series Editor, via pdavim@ua.pt.

Habeeb Adewale Ajimotokan

Principles and Applications of Tribology

Habeeb Adewale Ajimotokan
Department of Mechanical Engineering
University of Ilorin
Ilorin, Nigeria

ISSN 2191-530X   ISSN 2191-5318   (electronic)
SpringerBriefs in Applied Sciences and Technology
ISSN 2365-8223   ISSN 2365-8231   (electronic)
Manufacturing and Surface Engineering
ISBN 978-3-031-57408-5   ISBN 978-3-031-57409-2   (eBook)
https://doi.org/10.1007/978-3-031-57409-2

This Springer imprint is published by the registered company Springer Nature Switzerland AG
The registered company address is: Gewerbestrasse 11, 6330 Cham, Switzerland

Paper in this product is recyclable.

This research work is dedicated to the Supreme Being, the Almighty, for sparing my life up to this day and seeing me through every moment of my life. Also, to my dear parents, who gave me a love of life; my wife, who gave me a life of love; and my children, who gave joy and meaning to it all.

Preface

This book, a maiden edition, entitled *Principles and Applications of Tribology* offers in-depth knowledge of tribological concepts and how human daily life is impacted by tribological interactions. Its main goal is to assist in understanding the tribological principles and proper application, which is a necessary prerequisite for the best possible tribosystem design, operation and maintenance. The knowledge of tribology is significant because it deals with the principles of friction, lubrication and wear and their respective studies and applications; material and lubricant selections; as well as capacity determination; aids in the improvement of service life, safety and reliability of rubbing or sliding surfaces; and yields significant economic advantage. Its concepts are widely employed in the design of mechanical systems such as bearings, gears and internal combustion engines, as well as biomechanical systems such as human joint implants, among others. Also, they are applied in virtually every aspect of present and emerging technology, even in unanticipated realms like make-up or cosmetics applications.

All engineering students, regardless of their specialisation, can benefit from using the book, which was designed as a research text for senior undergraduate and graduate engineering students studying tribology. Additionally, engineers, academics, researchers in mechanical, biomechanics, and materials science and engineering to mention a few, as well as a host of other engineering professionals in allied industries, may find it to be a useful reference. The book covers a variety of topics associated with the multidisciplinary nature of modern tribology, such as the study of solid mechanics, which is necessary to estimate the contact stresses developed under asperity interactions, the study of fluid mechanics, which is necessary to determine the behaviour of lubricants and the formation of tribofilms between rubbing or sliding surfaces, and the study of material science and engineering, which focuses on atomic and microscale mechanisms that cause degradation or alteration of surfaces during relative motion. It is packed with research activities and more than a decade of tribology classroom instruction for senior undergraduate students and engineering professionals.

Among its many distinct features, the book offers unique basic engineering perspectives that are used to explain the concept of tribology and tribological interactions; basic principles of lubrication in minimising friction and protecting against wear of rubbing or sliding surfaces in relative motion; and practical, step-by-step instructions on how to choose the best lubricants and materials for tribological applications. By succinctly and clearly presenting the necessary concepts, the book seeks to bridge the gap between the common experience-based industrial practises and the scientific and engineering foundation of solution strategies for tribological challenges. One of the purposes of the book is to inspire learners to learn more about green tribology, as well as tribotronics, biotribology and tribology in nano- and space systems. The introduction and basic concepts of tribology; principles of friction, lubrication and wear; viscosity of fluid lubricants; surface studies and functionality characterisation; and lubricants and materials for tribological applications are the topics covered.

Ilorin, Nigeria   Habeeb Adewale Ajimotokan, (Ph.D.)

Acknowledgements

Above all, I give glory to the Almighty for granting me the opportunity and means to write and publish this book.

Many thanks go to Profs. I. K. Adegun, K. R. Ajao and Dr. P. O. Omoniyi of the Department of Mechanical Engineering, University of Ilorin; Dr. J. A. Adebisi of Department of Materials and Metallurgical Engineering, University of Ilorin; Dr. W. A. Jimoh of the Department of Fisheries and Aquaculture, University of Ilorin; Engr. M. A. Aladodo of the Department of Mechanical Engineering, Kwara State University; and my senior undergraduate students at University of Ilorin, Kwara State University and Elizade University and allied engineering professionals, who have provided insights during the book-writing processes.

The author is equally grateful to Prof. J. K. Odusote, Dean, Faculty of Engineering and Technology, University of Ilorin; Prof. A. S. Adekunle, Head, Department of Mechanical Engineering, University of Ilorin; Dr. K. O. Abdulrahman, Postgraduate Programmes Coordinator of the Department of Mechanical Engineering, University of Ilorin; Dr. A. B. Rabiu, Undergraduate Final Year Project Coordinator of the Department of Mechanical Engineering, University of Ilorin; and other colleagues in the department, faculty and the university at large for their various support during the writing and production processes.

Ilorin, Nigeria   Habeeb Adewale Ajimotokan, (Ph.D.)

About This Book

This book entitled *Principles and Applications of Tribology* provides a detailed knowledge of the concepts of tribology and how tribological interactions affect the daily lives of humans. The topics covered include the introduction and basic concepts of tribology; principles of friction, lubrication and wear; viscosity of fluid lubricants; surface studies and functionality characterisation; and lubricants and materials for tribological applications. The book, written as a research text in particular for senior undergraduate and postgraduate engineering learners studying the subject of *tribology*, can be used by engineers across all engineering disciplines. Also, it could serve as a useful reference for mechanical, biomechanics, and materials science and engineering researchers, engineers and academics and numerous other professionals in allied engineering industries.

The unique features of the book are its provision of distinct engineering perspectives used to outline and explain the concept of tribology and tribological interactions; discussion of the study and principles of friction, lubrication and wear; outline and discussion of the fundamental principles of lubrication for minimising friction and protecting against wear of lubricated contacts in relative motion, e.g., in mechanical systems such as bearings and gears, and machine elements, among others; discussion of the significant properties of commercial fluid lubricants; provision of pragmatic, step-by-step instructions on how to select the best lubricants and materials for tribological applications; and offer of outstanding service as a reference resource to academics and those in the research and development and industry; among others.

Contents

About the Author

Habeeb Adewale Ajimotokan (Ph.D., REng) is an Associate Professor and former Postgraduate Programmes Coordinator at the Department of Mechanical Engineering, University of Ilorin, Ilorin, Nigeria. He has been teaching or co-teaching tribology for over a decade. He has authored or co-authored several scientific articles, which comprise books, book chapters, refereed journal articles, edited conference papers and technical notes and/or reports in reputable outlets of international standard; reflecting his scholarly prowess, contributions to contemporary research, knowledge and consultancy services.

Chapter 1
Introduction and Basic Concepts of Tribology

Abstract The objectives of this chapter are to: (i) Define the term tribology; (ii) Outline the significance of the knowledge of tribology in developing mechanical systems such as bearings and gears, machine elements and numerous biomechanical systems; (iii) Outline and explain the concept of tribology and tribological interactions; (iv) Identify, outline and discuss the fundamentals of tribology; (v) Define the term tribosystems; (vi) Outline and describe the subfields of tribology; and (vii) Outline and discuss the multidisciplinary nature of tribology.

Keywords Tribology · Tribosystems · Biotribology · Nanotribology · Space tribology · Tribotronics · Tribological applications

1.1 Introduction

Every science has its own associated, unique concepts, terminologies and or nomenclatures whose precise respective definitions form a sound foundation for the development of the science and also help to prevent possible misconceptions. Tribology is no exception, and the term '*tribology*,' which best describes the earliest efforts to investigate the principle of rubbing or sliding surfaces, stems from the terms expressed in Greek as '*tribo*', that means '*rubbing*' and '*logy*', that means '*a branch of learning*', 'the *study of*' or '*knowledge of*' [1–3]. The same term, '*tribology*' is now widely used to refer to the scientific and technological fundamentals of friction, lubrication and wear, which are essential for the effective operation of many mechanical systems, including internal combustion engines; machine elements and biomechanical systems such as human joint implants and manufacturing, among others.

1.2   Tribology

Tribology is defined as the engineering science and technology of rubbing or sliding surfaces in relative motion, comprising the principles of friction, lubrication and wear and their respective studies and applications [1–4]. These two aspects of tribology, i.e., the engineering science—involves the basic mechanisms of tribosystems while the technology—involves the design, manufacture and maintenance of tribosystems and their applications, among others. The *knowledge of tribology* is becoming increasingly significant in the development of mechanical systems such as bearings, gears and internal combustion engines, numerous machine elements as well as biomechanical systems such as human joints, among others, because it deals with the *principles of friction, lubrication* and *wear* and their respective studies and applications, *material* and *lubricant selections,* as well as *capacity determination,* aids in the *improvement of safety, reliability and service life* of rubbing or sliding surfaces and yields *significant economic advantages* [5].

The *concepts of tribology* are widely employed in the design of mechanical systems such as bearings, gears and internal combustion engines, machine elements and biomechanical systems such as human joint implants, among others. Also, they are applied in virtually every aspect of present and emerging technology, even in unanticipated realms like make-up or cosmetics, including lipstick, powder and lip gloss applications [6]. Tribology plays a key role in manufacturing processes, like metal-forming operations, where *tool wear* together with the *required power* to machine a workpiece is increased by *friction.* These frictional effects result in not only an increased cost because of more frequent tool replacements, but also tolerance loss due to the often shift in tool dimensions and the requirement of greater forces for machining the workpiece. However, with the application of lubricant, a *layer of formed lubricating film* gets rid of the actual contacts of surfaces, resulting in the reduction of tool wear as well as the power needed to machine the workpiece by a third [7].

The *tribological interactions* of exposed solid surfaces with interacting substances and the surrounding might cause surface material losses and the *process of this loss of material* from such rubbing or sliding surfaces is termed '*wear*'. This wear might be minimised through *modification of the solids' surface characteristics* using one surface finishing or surface engineering technique or more, or through the *use of lubricants* [8]. The engineered surface increases the operational lifespan of original, recycled and resurfaced equipment, resulting in significant financial savings as well as material, energy and environmental conservation [6]. Any interaction between bearings, gears, machine elements or other mechanical and biomechanical systems where one slides, rubs or interfaces with another is influenced by the complexities associated with their interactions, whether lubricated (such as limb and hip prostheses) or unlubricated (like high-temperature sliding wear due to exposed interfacing solid surfaces where traditional lubricants can never be employed). For such unlubricated situations, the compacted oxide layer formation that coats the interacting surfaces has been established to eliminate friction and provide the required protection against wear.

1.3 Fundamentals of Tribology

The object of study in tribology is called a *tribosystem*, i.e., the physical system that comprises at least two contacting bodies and any environmental factor that affects their interaction [2, 9, 10]. In lubricated tribosystems, contact stress creates *tribofilms*—films that are formed on stressed lubricated tribological surfaces, typically *solid surface films*, resulting from the lubricant components' chemical reaction or tribological surfaces' reaction [2, 5, 11].

There are *four basic elements of tribology*. They include surface contact with its surroundings, comprising lubrication and lubricants; force generation and its interfacial transmission; material response to interfacial force generated; and design of tribological systems. The speciality area of *'green' tribology*, also known as eco-friendly tribology emphasises the scientific and technological green features for friction, lubrication and wear of rubbing or sliding surfaces in numerous systems. These eco-friendly surface interactions are vital for material sustainability and energy efficiency, and they have an influence on present and future environmental sustainability. In modern times, *bio-, micro-* and *nanotribology* are increasingly becoming significant because frictional interactions in microscopic components are vital in developing novel products such as in microchip technology, biomechanics and, by extension, almost all present and future technological products.

1.3.1 Tribosystems

A *tribosystem* is any of a class of tribological systems that transforms inputs by disturbance variables and interactions between elements into outputs and loss variables [9, 10]. The examples of the key features of a tribosystem, i.e., *inputs* to be transformed include motion type, motion sequence, load, temperatures and loading time; *disturbance variables* include materials properties and geometrical characteristics; possible *outputs* include force, torque, momentum, motion, shaft work and material variables; and loss variables include friction, wear and tear.

1.3.2 Subfields of Tribology

There are four basic *subfields* or *branches of tribology*. These subfields are biotribology, nanotribology, space tribology and tribotronics.

1.3.2.1 Biotribology

Biotribology is a subfield of tribology that deals with the study of friction, lubrication and wear of biotic or bionic systems [4, 12]. Generally, these processes in biological systems, including human joints like knees and hips, are examined using a tribological and contact mechanics framework on the basis that they exhibit a wide range of relative sliding and frictional interactions [1, 12]. As such, it is possible to consider both the tribological processes that naturally take place in or on the tissues and organs of biological systems as well as those that follow artificial implants.

1.3.2.2 Nanotribology

Nanotribology is a subfield of tribology that deals with the nanoscale study of friction, lubrication, wear and adhesion phenomena, where the effects of atomic and quantum interactions are significant [13]. The predominant objective of nanotribology subfield is to characterise and modify surfaces for purposes of scientific research and technological development through direct and indirect methods [14]. Direct methods such as microscopy techniques, comprising scanning tunnelling microscopes and atomic-force microscopes, among others, have been employed for analysis of extremely high-resolution surfaces, while indirect methods like computational approaches have also been comprehensively used [15, 16]. With the ability to change at the nanoscale the surface topology, friction might either be diminished or improved very thoroughly or vigorously than macroscopic lubrication and adhesion, and in this manner, superlubrication and superadhesion may be attained [13].

1.3.2.3 Space Tribology

Space tribology is a branch of tribology concerned with the study of tribosystems used in spaceship applications [16]. The goal of space tribology is to create dependable tribosystems that can resist the severe conditions of space.

1.3.2.4 Tribotronics

Tribotronics is a branch of tribology concerned with the study of interaction between triboelectricity and semiconductors. It uses the triboelectric potential to regulate semiconductors' electric conveyance and transformation (in sensing and actively controlling information; info-tribotronics) and semiconductors to regulate transmission and transformation of triboelectric power in circuits (in managing power and effective utilisation; power-tribotronics) [17].

1.4 Multidisciplinary Nature of Tribology

The multidisciplinary nature of modern tribology makes it a multidisciplinary engineering science and technology that involves knowledge based on other disciplines like:

(i) *Chemistry*—a subfield of science concerned with the substances of which matter is composed and their properties, e.g., in determining the composition of lubricant and how it's made up, the reactivity between lubricants, additives and solid surfaces, etc.;

(ii) *Physics*—a subfield of science concerned with the nature and properties of matter and energy, e.g., in establishing the structure of atoms, etc.;

(iii) *Solid mechanics*—a subfield of continuum mechanics concerned with the behaviour of solid materials under the mechanics of forces, temperature and phase changes and other internal or external agents; e.g., in estimating the contact stresses developed during the interaction of surface asperities, etc.;

(iv) *Fluid mechanics*—a subfield of engineering science concerned with the mechanics of forces and flow within fluids, e.g., in establishing lubricant behaviour and tribofilm formation between rubbing or sliding surfaces, etc.;

(v) *Material science and engineering*—a syncretic discipline that hybridise metallurgy, ceramics, solid-state physics and chemistry, e.g., in determining the structure of atoms, choice of materials and lubrication; explaining atomic and microscale mechanisms that cause degradation or alteration of surfaces during relative motion, etc.;

(vi) *Metallurgy*—a subfield of engineering science and technology concerned with the properties of metals, and their purification, production and possible usage, e.g., in selecting the choice of metallic material, solid lubrication, etc.;

(vii) *Metrology*—the scientific study of measurement e.g., for characterising surface irregularity, surface asperities, etc.; and

(viii) *Mass and heat transfer* e.g., surface or convective heat, etc.

1.5 Summary

Tribology is the engineering science and technology of rubbing or sliding surfaces in relative motion. The knowledge of tribology is significant in developing mechanical systems like bearings and gears, several machine elements and biomechanical systems such as human joint implants because it deals with the principles of friction, lubrication and wear; material and lubricant selections, as well as capacity determination; aids in the improvement of safety, reliability and service life of surfaces; and yields significant economic advantages. Its concepts are commonly employed in the design of bearings, gears, machine elements, human joint implants and almost every other aspect of modern technology.

A tribosystem is any of a class of systems that transforms inputs by disturbance variables and interactions between elements into outputs and loss variables.

The subfields of tribology include biotribology—the study of friction, lubrication and wear in biotic systems; nanotribology—the nanoscale study of friction, lubrication, wear and adhesion phenomena; space tribology—the study of tribosystems used in spaceship applications; and tribotronics—the study of interaction between triboelectricity and semiconductors.

References

1. Wang, O. J., & Chung, Y. (Eds.). (2013). *Encyclopedia of tribology*. Springer.
2. Abdelbary, A., & Chang, L. (2023). *Principles of engineering tribology: Fundamentals and applications*. Academic Press. https://doi.org/10.1016/B978-0-323-99115-5.01001-X
3. SynLube. (2007). *Tribology*. Retrieved from http://www.synlube.com/tribolog.htm
4. Davim, J. P. (Ed.). (2013). *Biotribology*. Wiley.
5. Hirani, H. (2016). *Fundamentals of engineering tribology with applications*. Cambridge University Press.
6. Britannica, T. (Ed.). (2010). Lubrication. *Encyclopedia Britannica*. Retrieved from https://www.britannica.com/technology/lubrication
7. Ajimotokan, H. A., & Mahamood, R. M. (Eds.). (2017). *Engineering workshop technology*. University of Ilorin Publishing House.
8. Chattopadhyay, R. (2004). *Advanced thermally assisted surface engineering processes*. Kluwer Academic Publishers.
9. Mang, T., Bobsin, K., & Bartels, T. (2011). *Industrial tribology: Tribosystems, friction, wear and surface engineering, lubrication*. Wiley-VCH Verlag & Co.
10. Blau, P. J. (2017). *Tribosystem analysis: A practical approach to the diagnosis of wear problems*. CRC Press, Taylor & Francis Group.
11. Pawlak, Z. (2003). *Tribochemistry of lubricating oil*. Elsevier.
12. Ostermeyer, G, Popov, V. L., Shilko, E. V., & Vasiljeva, O. S., (Eds.). (2021). *Multiscale biomechanics and tribology of inorganic and organic systems*. Springer Tracts in Mechanical Engineering (a Part of Springer Nature Switzerland AG).
13. Kandile, N. G., & Harding, D. R. K. (2014). Nanotribology: Progress towards improved lubrication for the control of friction using ionic liquid lubricants. In G. Biresaw, K. L. Mittal (Eds.), *Surfactants in tribology* (Vol. 4). CRC Press, Taylor & Francis Group.
14. Bhushan, B. (Ed.). (2020). *Handbook of micro/nanotribology* (2nd ed.). CRC-Press, Boca Raton.
15. Bhushan, B., Israelachvili, J. N., & Landman, U. (1995). Nanotribology: Friction, wear, and lubrication at the atomic scale. *Nature, 374*(6523), 607–616.
16. Davidson, E. (2022). *Introduction to tribology*. States Academic Press.
17. Xi, F., Pang, Y., Li, W., Jiang, T., Zhang, L., Guo, T., Liu, G., Zhang, C., & Wang, Z. L. (2017). Universal power management strategy for triboelectric nanogenerator. *Nano Energy, 37*, 168–176.

Chapter 2
Principles of Frication, Lubrication and Wear

Abstract The objectives of this chapter are to: (i) Define term friction and specify and describe its types; (ii) Outline and describe the ways of reducing friction between rubbing or sliding surfaces; (iii) Identify and outline the rules of sliding friction; (iv) Identify, outline and describe coefficient of friction and its measurement; (v) Define the term lubrication and outline its objectives; (vi) Identify, outline and briefly explain factors that affect lubrication; (vii) Outline and highlight the fundamental principles of lubrication to minimise friction and protect against wear, e.g., of mechanical systems such as bearings and gears, machine elements and biomechanical systems, among others; (viii) Identify, outline and discuss varieties of lubrication; (ix) Identify, outline and discuss types of lubricating films; (x) Identify, outline and describe lubrication regimes; (xi) Define the term wear and highlight the working environmental factors that affect it; and (xii) Identify, outline and discuss classifications of wear modes.

Keywords Friction · Dry friction · Fluid friction · Lubrication · Lubrication regimes · Lubricated contacts · Wear

2.1 Introduction

The efficiency of machine elements, mechanical and biomechanical systems as well as their service lives, might all be impacted by the complicated, interwoven subjects: *friction*, *lubrication* and *wear* [1]. Despite the importance of all three, lubrication and wear are given the most attention in tribology. Friction is undesirable in most operating machines because it results in *energy dissipation or loss* and *performance deterioration* due to heat generation. As a result, efforts are made to minimise friction by *employing low-friction materials, surface films and coatings* and *surface modification or redesigning to protect against wear* [2]. Furthermore, the speciality area of '*eco-friendly*' tribology underscores the green technology aspects of friction, lubrication and wear of opposing solid contacts in relative motion in some mechanical and biomechanical systems and almost every other aspect of modern technology.

2.2 Friction

Friction is a force due to the tangential resistance to motion that resists the relative rolling or sliding of solid surfaces and material elements against each other [3]. Friction is caused by the dissipation of energy in the rubbing or sliding interface, which results in the generation of heat. The generated heat has a negative impact on the majority of system operations and prevents free rubbing or sliding at such interfaces, which has a significant impact on the flow and deformation of materials, such as in metalworking operations. Friction is also not always detrimental because it often plays a crucial role in the operation of numerous mechanisms. For example, without friction, it would be practically impossible to hold drill bits in the chucks or jaws of machine tools while they were being machined or manufactured, roll metal, or clamp workpieces on machine tools. A *vehicle's braking coefficient* depends on the surface friction between a vehicle's wheels and the pavement surface and accordingly, less friction results in a lower vehicle braking coefficient and braking response [4].

High friction is necessary for the satisfactory functioning of paper clips, tongs, nuts and bolts, and for everyday activities like walking and manual grasping of objects [3]. Walking without slipping requires frictional forces, such as traction, which may be advantageous but also present a lot of resistance to motion. For instance, an automobile's engine uses about 20% of its power to overcome the frictional forces in its moving parts. But for items like engines, skis and watch internal mechanisms that are meant to move continuously, low friction is preferred while brakes and clutches need constant friction to prevent the unpleasant jerky movements that would otherwise occur [4].

The *factors that affect friction* between rubbing or sliding surfaces include the presence of wear and external particles on the bounding interface; the relative hardness of the materials in contact; external loads or displacements; environmental conditions like temperature and lubricants; surface topography; the microstructure or morphology of the materials; the apparent contact area; and the kinematics of the surfaces in contact (i.e., the direction and magnitude of the motion) [5]. Adhesion, mechanical contacts of opposing surface asperities, deformation and or fracture of contact surface layers, and primarily agglomerated wear particles trapped between the moving surfaces are some of the microscopic mechanics involved in generating friction [6].

The phenomenon known as friction has numerous diverse explanations, including the adhesion and abrasion theories of friction, which are, among them, widely accepted. The adhesion theory is based on the observation that two clean, dry surfaces only make contact with one another at a small portion of their apparent contact area, regardless of how smooth they are. The surface's maximum slope typically ranges from 5° to 15° and in this case, tiny asperities—which are irregularities or small projections from the surface—that are in contact with one another—support the normal (contact) load [7, 8]. Because of this, the asperities' normal stresses are high, leading to plastic deformation at the junctions, and as a result of their contact, microwelds are formed by the asperities [8]. The abrasion theory, however, is predicated on the idea that an asperity from a hard surface such as a tool or die—penetrates and ploughs through a softer surface such as a machined workpiece. Similar to cutting and abrasive processes, ploughing may result in material displacement as well as small chips or slivers.

The majority of the energy lost in overcoming friction is converted into heat, raising the surface temperature. A very small portion of the energy is stored on surfaces that have undergone plastic deformation. This surface temperature rises as sliding speed and friction increase, thermal conductivity declines and the specific heat of the materials being moved increases [9]. The temperature at the interface may be high enough to cause surface softening or melting, and occasionally, can alter the microstructure of the materials. Note that the viscosity and other characteristics of lubricants are also influenced by the surface temperature, with sufficiently high temperatures leading to their breakdown. As a result, friction control is crucial, with the main objectives being to broaden the stage of the stable wear and minimise the wear and surface damage intensities during the initial (or running-in) stage of the wear process, which may be accomplished through surface engineering [9].

2.2.1 Types of Friction

In general, there are primarily *four types of friction* that are often encountered in practice and considered in tribology [9–11]. These friction types are dry friction, fluid friction, mixed friction and boundary friction. Attempts to improve energy efficiency of mechanical systems such as bearings and gears, and machine elements have resulted in the use of lower-viscosity lubricants and a higher likelihood of mixed or boundary friction [12].

Dry friction, also called Coulomb friction, is the tangential component of a contact force that opposes the developed lateral motion when two dry surfaces tend to slide in relation to one another [13]. Apart from friction between atoms or molecules, dry friction typically occurs when two surfaces slide against one another without any extra lubricants or layers due to the contact of opposing surface asperities. Dry friction can be categorised into *kinetic friction*, which occurs between opposing moving surfaces and *static friction* (also called stiction), which occurs between stationary surfaces; *fluid friction*, which is the tangential component of a contact force that occurs between adjacent layers of a viscous fluid when those layers are moving in relation to one another [13]. Fluid friction depicts the friction that occurs when layers of a viscous fluid move in relation to one another. Lubricated mechanisms experience fluid friction, which is the result of the presence of a lubricant between lubricated sliding or rubbing surfaces such as in a fluid between surfaces of bearings, gears or machine elements [10]; *mixed friction*, which is the region where the dry and fluid friction coexist in the contact zone [11]; and *boundary friction*, which often develops during sliding friction when a lubricant thickness is thinner than ten atomic layers or two rubbing surfaces that are separated by a lubricant layer make contact with one another because of surface asperities [11, 14].

2.2.2 Reducing Friction

Friction can be minimised between rubbing or sliding surfaces through the *selection of low-friction materials*, i.e., selecting appropriate low-friction materials with low adhesion, such as carbides and ceramics; *using surface films and coatings* to minimise adhesion and interactions of the bounding surface with the other, i.e., deploying lubricants like oils, air and steam that interpose fluid films, or graphite and molybdenum disulphide (MoS_2) that interpose adherent solid films between sliding surfaces. The lubricant creates surface films during rubbing or sliding of the surface, minimising friction and protecting against wear; and *subjecting the rubbing or sliding surfaces to ultrasonic vibrations*, generally at 20 kHz [15]. The vibrations' amplitude periodically separates the rubbing or sliding surfaces to permit a more spontaneous flow of lubricant into the rubbing or sliding interface during these separations.

2.2.3 Rules of Sliding Friction

Generally, over a broad range of applications, two fundamental laws of intrinsic (or conventional) friction are obeyed. Despite having been first described by Leonardo da Vinci some 200 years earlier, these rules are commonly known as Amontons equations, after Guillaume Amontons—a French physicist who rediscovered the rules in 1699. According to the first sliding friction rule, friction force and nominal load are directly proportional [16]. Mathematically, the first rule can be defined using Eq. 2.1:

$$F = \mu W, \tag{2.1}$$

where F denotes the friction force, μ is a proportionality constant termed the coefficient of friction, which can be the coefficient of static friction, μ_s, or kinetic friction, μ_k and W is the nominal load. Equation 2.1 is independent of the normal load.

On the other hand, it is convenient to often state this rule using the frictional angle or constant angle of repose. Thus, the second rule of sliding friction states that the coefficient of static friction is equivalent to the tangent of frictional angle [16]. Mathematically, the second rule can be defined using Eq. 2.2:

$$\mu_s = \tan \theta, \tag{2.2}$$

where θ denotes the tangent of frictional angle or constant angle of repose. Equation 2.2 depicts the frictional angle, θ, as the angle at which a body of any weight will remain stationary when positioned on a plane inclined at a less than frictional angle θ from the horizontal [7]. If the inclination angle is increased to θ, however, the body would begin to slide downward.

2.2.4 Coefficient of Friction

The coefficient of friction, a dimensionless quantity, is a constant ratio that has been found through experimentation to exist between the normal reaction between two surfaces and the magnitude of limiting friction. The application of a frictional force, also known as a tangential force, is necessary for the sliding movement between two bodies with an interface [17]. This force, known as the friction force, is necessary to push through the softer material or shear the junctions. The coefficient of friction can be employed to determine the force of friction. Therefore, the *coefficient of friction* is defined as the ratio of the friction force between two surfaces in contact with the existing normal force that exists between the object resting on the surface and the surface [16, 18]. Mathematically, the coefficient of friction, μ, can be expressed using Eq. 2.3:

$$\mu = \frac{F}{N},\tag{2.3}$$

where F denotes the friction force and N is the normal force. In particular, but not only, the two surfaces' physical properties, the point of contact's ambient temperature, and the object's speed across the surface would all affect this ratio. The amount of surface friction that is present on a particular vehicle at a given time is directly proportional to the amount of braking applied, with the exception of when wheel lockup and antiskid protection systems that are present on the majority of contemporary transport vehicles—are engaged [16]. Bounding surfaces are assumed to be either rough (i.e., tangential forces prevent relative motion between surfaces) or frictionless (i.e., surfaces could move freely with respect to one another) [17]. In essence, *no surface can be completely frictionless* because in any given interaction between two surfaces, tangential forces—also known as friction forces—would often arise whenever one surface tried to move in relation to the other. Though friction forces have a limited magnitude, they would not impede motion if large enough forces were applied. Thus, the difference between rough and frictionless is a matter of degree.

2.2.5 Friction Measurement

Though the coefficient of friction might be computed theoretically, it is typically found through experimentation, either with small-scale specimens of various shapes in simulated laboratory tests or during actual manufacturing operations. The ring-compression test is a widely accepted test, in particular for bulk deformation operation [7, 19]. Between two flat platens, a flat ring is agitated plastically. Radially outward, the ring expands as its height decreases. The ring's inner and outer diameters expand in a similar way as a solid disc if there is no friction at the interfaces. By measuring the specimen's change in internal diameter and using the curves (see Fig. 2.1) that resulted from theoretical analyses, the coefficient of friction can be ascertained. Keep in mind that every ring geometry and material has a unique set of curves. A specimen with an outer diameter, inner diameter and height proportion of 6:3:2 is considered to have the almost usual geometry. In these tests, the actual size of the specimen usually does not matter. The magnitude of the friction coefficient can be directly read from this chart (i.e., Fig. 2.1) once the percentage reduction in the internal diameter and height of the ring has been established [20].

Example 2.1 When a ring specimen experiences a 40% reduction in height and a 10% decrease in internal diameter, calculate the resulting coefficient of friction.

Solution

Given: Reduction in height, $\Delta H = 40\%$; Reduction in internal diameter, $\Delta ID = 10\%$ and Coefficient of friction, $\mu = ?$

Therefore, the coefficient of friction, μ, can be obtained from the ring-compression test chart (see Fig. 2.1) with a 40% reduction in height and computed 10% reduction in internal diameter.

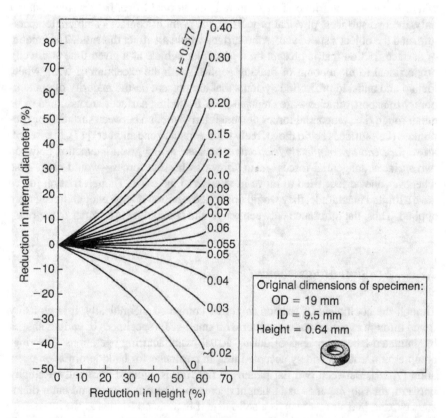

Fig. 2.1 Chart of a ring-compression test to determine friction coefficient. *Source*: (Kalpakjian and Schmid, [19])

Using the ring-compression test chart, the resulting coefficient of friction $\mu = 0.10$.

Example 2.2 A ring-compression test specimen of 10 mm height with inside and outside diameters of 15 mm and 30 mm, respectively is reduced in thickness by 50%. If the outside diameter is 38 mm after deformation, determine the.

(i) New inside diameter and
(ii) coefficient of friction.

Solution

Given: Initial height, $H_o = 10$ mm; Initial inside diameter, $ID_o = 15$ mm; Initial outside diameter, $OD_o = 30$ mm; Reduction in height, $\Delta H = 50\%$; and New outside diameter, $OD = 38$ mm.

(i) New inside diameter, $ID = ?$

Therefore, the new ID can be determined using volume constancy, i.e.,

$$\frac{\pi}{4}\left(OD_o^2 - ID_o^2\right)H_o = \frac{\pi}{4}\left(OD^2 - ID^2\right)H$$

$$\frac{\pi}{4}\left(30^2 - 15^2\right)10 = \frac{\pi}{4}\left(38^2 - ID^2\right)5 \text{ where } H \text{ is } 50\% \text{ of } H_o.$$

New inside diameter $ID = 9.7$ mm.
Hence, the change in internal diameter, $\Delta ID = \frac{ID - ID_0}{ID_0}$

$$= \frac{9.7 - 15}{15}$$
$$= -0.35 \text{ or } 35\% \text{ (reduction)}.$$

(ii) Coefficient of friction, $\mu = ?$

Therefore, the coefficient of friction, μ, , can be obtained from the ring-compression test chart (see Fig. 2.1) with a 50% reduction in height and computed 35% reduction in internal diameter.

Using the ring-compression test chart, the resulting coefficient of friction $\mu = 0.21$.

Exercise 2.1
Use Fig. 2.1 for the ring-compression test chart.

1. When a ring specimen experiences a 50% reduction in height and a 30% decrease in internal diameter, determine the resulting coefficient of friction.
2. A specimen of a ring-compression test has a height of 10 mm and inside and outside diameters of 15 mm and 30 mm, respectively is reduced in height by 45%. If the outside diameter was 38 mm after deformation, determine the

  (i) New inside diameter and
  (ii) Coefficient of friction.

2.2.6 Areas of Applications

In several engineering disciplines, friction plays an important function in both the design of contemporary engineering systems and in everyday activities. Therefore, it is essential to have a good understanding of the principles of friction. In transportation, friction plays a vital role. Automobile brakes are based on friction by nature; they slow a car by turning its kinetic energy into heat. While designing brake systems, one of the technological challenges is, incidentally, safely dispersing this significant amount of heat. The disc and brake pads that are compressed transversely against the rotating disc generate friction, which is the basis for disc brake action. Brake shoes or pads are forced outward against a revolving cylinder (i.e., brake drum) in drum brakes to produce friction [21]. Disc brakes provide superior stopping power over drum brakes because they can be cooled more effectively.

Road slipperiness is a crucial component of automobile design and safety. The way tyres interact with the driving surface is influenced by road roughness. On a broader scale, friction plays a major part in the development of measuring instruments such as the tribometer, which measures friction on a surface, and the profilograph, which gauges the roughness of pavement surfaces [21]. Friction finds wide-ranging uses in households. Heat and ignition of matchsticks are achieved through friction, which is the result of contact between the matchstick's head and the rubbing surface of the matchbox. By significantly raising the coefficient of friction between the object and the surface, sticky patches help keep objects from slipping off smooth surfaces.

2.3 Lubrication

Lubrication is the process or technique used to interpose a substance referred to as lubricant between two surfaces to carry the load or cushion the pressure generated by the opposing surfaces in close proximity to minimise friction and protect against wear of one or both surfaces [22, 23]. Most of the lubricants that are interposed are liquids, such as petroleum or mineral oils, silicone fluids and synthetic esters. However, they can also be solids, like graphite and molybdenum disulphide (MoS_2), or solid–liquid or solid–liquid dispersions, such as greases, which are used in rolling element bearings; or exceptionally gaseous, such as air, industrial gases and liquid–metal vapours, which are used in gas bearings [24]. The applied load or pressure generated by bounding surfaces is typically carried by the pressure that is generated within the lubricating fluid as a result of the fluid's viscous frictional resistance to motion between the surfaces [24, 25].

The machine elements that need to be lubricated are wire ropes; flexible couplings, chains, cam followers and cams; rolling elements; gears; cylinders; and slides, guides and ways. These components have fitted or formed surfaces that can move in combinations of these ways, such as sliding, rolling, advancing and retreating, with respect to one another. High frictional forces that result in high temperatures and exposure of interfacing surfaces to wear or possibly failure take place if there is actual contact between the surfaces. As a result, mechanical systems such as bearings and gears, and machine elements, among others are lubricated to minimise or eliminate actual surface contact. Most machines would only operate for a short periods of time without lubricant. The most serious *consequence of inadequate lubrication* is typically excessive wear because, eventually, the machine's components—bearings, gears, or other moving parts—will become unusable and must be taken out of service for repair or replacement. This usually happens after a brief period of operation. Repair or replacement expenses, including labour and materials, might be high, but the biggest expense could be in lost output or machine downtime [26].

The frictional forces between surfaces could be so high in cases of insufficient lubrication that drive motors would be overloaded or there would be excessive frictional power losses even before elements fail from excessive wear. Furthermore, poorly lubricated machinery would not function effectively, quietly, or smoothly. Bearings, gears and machine elements are lubricated by fluid films that are positioned and maintained between moving surfaces during lubrication. Because shearing happens easily, these films minimise or eliminate the actual contact between the surfaces and minimise the frictional force preventing the surfaces from moving [24].

2.3.1 Objectives of Lubrication

The predominant objectives of lubrication are to [24, 25, 27]:

(i) Form elastohydrodynamic (EHD) lubrication films at the rolling contacts between the rolling elements and the raceways, or thin films between sliding surfaces. These films may maintain complete surface separation of rolling elements and raceways or sliding surfaces under suitable operating conditions of viscosity, speed and load, significantly extending the life of mechanical systems, or machine elements;

(ii) Minimise friction and protect against wear in applications with inadequate EHD or thin-film lubrication. Past experiences have shown that mechanical systems like rolling bearings and gears, and machine elements, among others function well under mixed lubrication conditions for extended periods of time if the right lubricant is applied. Perfect circumstances of total separation are not always upheld in real life. Surface asperities come into contact and mixed friction, or hydrodynamic coupled with direct contact friction, occurs if the height of the asperities is greater than the formed EHD or thin lubrication film; and

(iii) Cool mechanical systems such as bearings and gears, and machine elements and minimise the maximum contact temperature of the rolling elements and the raceways or the sliding surfaces (as applied to fluid lubricants). Enough lubricant circulation to dissipate heat from mechanical systems is necessary for effective cooling. Running the oil through an external heat exchanger provides the best cooling. A basic oil sump system, however, can improve the heat transfer from mechanical systems, or machine elements by convection even in the absence of complex circulation. Greases and solid lubricants can only be used or restricted to relatively low-speed applications because they are ineffective at cooling.

Additional lubrication objectives include *vibration damping, corrosion protection* and *contaminants and wear debris removal* from the raceways or bounding surfaces (as applied to liquid lubricant) [23]. As a damper, a sufficient EHD or thin film is crucial, which effectively isolates vibrations by supporting shafts without making contact with them. To lessen noise and vibrations in mechanical systems, an appropriate EHD or thin film may be useful. Additionally, lubricants serve as shock-damping fluids in certain devices for energy transfer (such as shock absorbers) and around machine components like gears, which experience high, inconsistent loads or exerted pressures, and as electrical insulators—lubricants with high dielectric constants, used in specialised applications like transformers and switchgear [23]. Lubricants are required to be kept free of impurities and water to have the best insulating qualities.

2.3.2 Factors Affecting Lubrication

A number of variables determine whether grease or oil is employed for lubrication of mechanical systems, such as gears and bearings, and machine elements, among others. Machines with installed bearings often use the same oil that is used for other parts of the machine. Grease is often utilised; however in some cases, it might be more feasible to lubricate each bearing individually. Groups of similarly linked bearings can be lubricated using a *circulation system*, an *oil mist system*, or a *centralised lubrication system* [24]. The qualities of these various application methods have a multifaceted impact on the selection of oil or grease. Many contemporary manufacturer of bearings now offer them '*packed for life*' with grease, which implies that they will not need to be lubricated again for the duration of their service life.

The EHD lubricating film thickness generated between the lubricated contact areas of rolling elements and raceways, or thin-film thickness formed between lubricated sliding surfaces is a function of the *speed* at which the surfaces rub or slide relative to one another, *load (or pressure* exerted) by the opposing surfaces, *oil viscosity (or grease consistency), operating temperature* and *contamination* [25]. Increases in load (i.e., pressure exerted) or operating temperature cause the formed *film thickness* to diminish because oil viscosity decreases at higher temperatures. The formed film thickness increases with rolling or sliding surface speed or oil viscosity. To ensure a sufficient minimal film thickness, the oil viscosity should be appropriately chosen for each set of operating conditions. While the requirements of an EHD or thin lubricating films are generally the key factor in lubrication choices, the lubrication requirements of a bearing's or machine element's other rubbing or sliding surfaces must also be taken into account [25]. The pressure needs to be strong enough to induce the contacting surfaces to deform elastically for EHD or thin lubricating film formation to take place.

2.3.2.1 Effect of Speed

Like oil viscosity or grease consistency, speed has a similar influence on the thickness of the lubricating film, for example, increasing the speed by two folds could lead to a lubricating film thickness increase of more than 50% [25]. The physical characteristics of the grease are the primary factor to be considered when reckoning with bearing speed for grease-lubricated bearings, gears or machine elements. The grease should be sufficiently soft to plummet gradually in the direction of the rolling elements at low-to-moderate speeds, but not so soft as to let excess material to get in the way of them [25]. Over grease can result in high operating temperatures due to an increase in shearing friction. Relatively stiff grease can be used for high speeds, but it should not be so stiff that the rolling components cannot pick up and distribute enough grease to continuously replenish the lubricating films after they have cut a channel through it. Furthermore, the grease needs to have a strong resistance to softening from mechanical shearing to maintain low shearing friction and stop leaks from housing seals.

2.3.2.2 Effect of Load

Unlike the effect of oil viscosity or speed, the load (i.e., pressure exerted) has a minor impact on the lubricating film thickness under EHD or thin lubricating film circumstances, e.g., increasing the speed by two folds may only lead to a 10% decrease in lubricating film thickness [25]. Consequently, in situations with consistent loads, the operating temperature and bearing speed factor may typically be used to determine the oil's viscosity without taking the load into account. Higher viscosity oils are typically needed when shock or vibratory loads are present because they inhibit surface asperities interactions through the film. Lubricants with improved antiwear qualities could be preferred under extreme shock loading circumstances.

2.3.2.3   Effect of Temperature

The operating temperature of mechanical systems, such as bearings and gears or machine elements, among others, must always be considered when selecting lubricants, as oil viscosity or grease consistency are functions of temperature [24, 25, 28]. Bearing operating temperatures can rise above normal due to heat transfer from a hot shaft or spindle or heat radiated to the housing from a heated environment. Bearing temperatures might also rise as a result of grease overfilling causing its excessive churning out. Higher than usual temperatures cause an oil's viscosity to decrease, which can lead to a drop in film thickness below a safe threshold and excessive spinning and additional frictional heating of the grease. The oxidation-related degradation of oil or grease is also accelerated by high temperatures [28]. Oil may thicken as a result of oxidation, and eventually, deposits may form that impede the operation of machine elements, bearings, gears, or oil flow. Additionally, oxidation can result in deposits containing greases and, in extreme circumstances, hardness that prevents the grease from feeding and lubricating [29]. To start any mechanical system at low temperatures, the lubricating fluid should be in a state that allows for adequate power distribution to protect against undue wear prior to the frictional heating that warms the system and its lubricant [24].

2.3.2.4   Contamination

The most common reason for reduced bearing or machine life is the presence of kind of solid particles wedged between rolling elements and raceways or sliding surfaces. As a result, it is important to keep dirt out of bearings, gears and machine elements wherever feasible (including during storage times before installation), and to replace lubricants prior to when oxidation would attain an advanced state of deposit formation [25]. The time duration that lubricating fluids might be left-in-service without experiencing undue oxidation can be significantly increased by using premium-quality lubricants that inhibit oxidation. Water incursions into mechanical systems or machine elements often shorten their fatigue lives and lead to corrosion, which can swiftly destroy the component or system [24, 25]. Grease that comes into contact with water has the potential to soften and leak from bearings, gears, or other machine components. Water with a lot of volume might remove the lubrication. Acidic fluids can occasionally seep into machine elements, gears, or bearings and cause corrosion. Any of these circumstances often call for extra safety measures and perhaps specialised lubricants [24].

2.3.3   Varieties of Lubrication

Generally, the *three basic varieties (or types) of lubrication* are fluid-film, boundary and solid lubrication [23, 28]. The best among them, which is the most desirable lubrication form, is the fluid-film lubrication because the films generated by the motion of lubricated contacts are sufficiently thick to completely separate the load-carrying surfaces under normal operating conditions [28]. Consequently, because friction is minimised to a practically minimum due to the shearing of the lubricating fluid films alone, there is essentially no mechanical contact of opposing surface asperities, and thus, the surfaces in contact are protected against wear.

2.3.3.1 Fluid-Film Lubrication

Fluid-film lubrication is a lubrication type in which a fluid film is interposed between sliding surfaces using a fluid-film lubricant to minimise friction and protect against wear [16, 28]. In fluid-film lubrication, the fluid-film lubricants can be introduced unintentionally—for example, water between a wet pavement and an automobile wheel—or purposefully—such as oil in an automobile's main bearings [9]. While the fluid-film lubricant is typically a liquid, it can also be a gas, with air being the most widely used type [23]. To keep the sliding surfaces apart, the pressure within the lubricating film needs to be higher than or at least balance the load on them. The surface is considered *hydrostatically lubricated* if the lubricating fluid's pressure that forms the lubricating fluid film is provided through an external source and *hydrodynamically lubricated* if the lubricating fluid's pressure that forms the lubricating fluid film is caused by a viscous drag resulting from the form and motion of the surfaces themselves [23, 24]. This lubricating process is dependent on the lubricant's viscosity.

2.3.3.2 Boundary Lubrication

Boundary lubrication can be defined as the state or condition that exists or lies between unlubricated and fluid-film lubrication sliding [23]. It is also the lubrication condition, where the lubricant's properties and the surfaces, as opposed to viscosity, control the friction between the surfaces [29]. Much of the lubrication phenomenon that often happens when machines start and stop constitutes a significant part of boundary lubrication [23].

2.3.3.3 Solid Lubrication

Solid lubrication is a lubrication type in which a solid film is interposed between two surfaces in contact using a solid-film lubricant to minimise adhesion and friction, and protect against wear [28, 30]. The solid-film lubricants, which are paint-like coatings made of tiny lubricating pigment particles mixed with a binder and other additives are applied to substrates through brushing, dipping, or spraying techniques, and after curing, they solidify into a film that decreases friction, repels water and lengthens the substrate's wear life [31]. These solid-film lubricants are available from a wide range of solid substances with inherent lubricating capabilities [32]. Graphite, polytetrafluoroethylene (PTFE), also referred to as Teflon, and MoS_2 are the most widely employed substances for solid lubrication among others, like perfluoroalkoxy (PFA) copolymers, fluorinated ethylene propylene (FEP), tungsten disulphide, boron nitride, lead oxide and antimony oxide [32]. No single formulation based on economic and comparative advantages, may fulfil all of the necessities for a specific solid lubrication. Properties that must be put into consideration besides the requirements and environmental conditions under which the solid-film lubricant should operate include the coefficient of friction, load-carrying capacity, corrosion resistance (or vulnerability to galvanic corrosion) and electrical conductivity [32–34].

Solid lubricants are primarily employed in situations when *extreme pressure* (EP), *antiwear type additives*, or *temperature* are required or when standard lubricants are *insufficiently resistant to load or temperature extremes* [23, 31]. The most widely employed application of solid lubricants, in particular graphite and MoS_2, is as an EP ingredient in grease formulations [33]. By making it easier for the surfaces in motion to glide over one another, the plate-like shape of solid lubricant particles lowers friction. Bushings and pins can both benefit from this application. Where moisture is present, graphite is, in particular helpful and in actuality, the complete friction-reducing potential of graphite depends on the presence of moisture [33].

2.3.4 Types of Lubrication Films

There are three types of lubrication (or lubricating) films [24, 25]. They are fluid, thin and solid films.

2.3.4.1 Fluid Films

Fluid film refers to the sufficiently thick films generated by the opposing motion of lubricated contact during normal operation that maintains complete surface separation [24]. There are three different ways by which fluid films can be formed. They are *hydrodynamic and EHD films*, which is a fluid film generated by the motion of lubricated contacts in which the lubricating fluid film's pressure is produced through a viscous drag due to the surfaces' shape and motion through a convergent zone. This ensures that the developed pressure within the film is sufficient to maintain surface separation; *hydrostatic film*, which is a fluid film formed by the motion of lubricated contacts in which the lubricating fluid film's pressure is supplied by a source externally between the surfaces that might or might not be moving relative to one another; and *squeeze (or mixed) film*, which is a fluid film formed by the motion of lubricated contacts in which the pressure of the lubricating film is produced through the movement of the opposing lubricated surfaces towards one another [23–25].

2.3.4.2 Thin Films

Thin film refers to the not sufficiently thick films formed during opposing motion of lubricated contacts that maintain complete surface separation [25, 28]. Maintaining fluid films requires an ample, constant supply of lubricant, and it is not often feasible or practical to lubricate bearings, gears, or other machine elements with such a quantity of lubricant. In other circumstances, the loads and speeds involved, for example, make it impossible to maintain fluid film when a hydrodynamic film bearing is initially activated. In these circumstances, lubrication takes place under what is termed '*thin film*'. In thin-film interactions between surfaces, there is frequently sufficient oil present to allow for a combination of contact between surfaces and fluid films to carry the load in part. This condition is commonly referred to as *mixed-film lubrication* [28]. With insufficient lubricant or as load increases, a threshold is attained where the fluid lubricant's ability to separate the surfaces decreases. This situation is frequently referred to as boundary lubrication.

2.3.4.3 Solid Films

Solid film, also known as dry film, refers to the film formed by more or less permanently bonded solid lubricant to the bounding surfaces [28]. A number of factors, including environmental conditions, sealing challenges and application difficulties, render oils and greases unusable in many applications. For these kinds of applications, several lubricating films that are bonded more or less permanently have been established to minimise friction and protect against wear. Through rubbing action or chemical interaction, the solid-film lubricants develop themselves on the surfaces and lower the effective surface roughness. A low-friction solid lubricant, like MoS_2, suspended in a carrier and employed essentially in a similar manner to a typical lubricant, is the most basic solid lubricating film type [30, 35]. When the carrier—which could be grease, volatile solvent, or any of a number of other materials—is squeezed, clenched or evaporated from the surfaces, it leaves behind a MoS_2 layer that acts as a lubricant [35]. Additionally, different resin types, cured for the formation of compactly adherent coatings with suitable frictional characteristics, are used to bond solid lubricants to rubbing surfaces, such as in some plastic bearings and sintered metal bearings [24]. A lubricating coating may then be formed on the surface in contact as a result of some of the solid lubricant being transferred during operation.

2.3.5 Lubrication Regimes

Based on the Stribeck curve for self-pressure generating lubricated contacts, there are four distinct lubrication forms known as *lubrication regimes* [9]. This is the *boundary lubrication regime*, which is a condition that develops when the fluid films of lubricated contacts are negligible with considerable contact of opposing asperities

[36]. As some of the surface asperities are higher than the mean film thickness, the coefficient of friction increases until it reaches a maximum; *mixed lubrication regime*, which is a condition that develops when the mean film thickness of lubricated contacts is slightly higher than the surface roughness. This causes the tallest asperities of the opposing contacts to sometimes protrude through the film and make contact. As a result, the coefficient of friction gradually decreases until a minimum is attained; *EHD lubrication regime*, which is a condition that develops when a lubricant is introduced between rolling surfaces in contact, such as ball and rolling element bearings. The load is sufficiently high in this lubrication regime to generate pressures that cause the lubricated contacts to elastically deform, which results in a low coefficient of friction in this area; and *hydrodynamic lubrication regime*, which is a condition that develops when a relatively thick film separates the load-carrying surfaces, offering a stable lubrication regime and preventing surface asperities from making contact during the bearing's steady-state operation [9, 37].

For external pressure-generating lubricated contacts, a lubrication regime in this category is called *hydrostatic lubrication regime*, which is a condition in which lubricated contacts are completely separated by a lubricating fluid film supplied by an external pressure [9, 24, 30]. Figure 2.3 shows the Stribeck curve, indicating the various lubrication regimes on the basis of their tribological characteristics [9]. The Stribeck curve depicts friction as a function of load, speed and viscosity. Its horizontal axis represents a dimensionless parameter that combines load, speed and viscosity, and its vertical axis plots the friction coefficient. According to Richard Stribeck, the Stribeck curve is essentially a curve between the coefficient of friction and a number defined by the dynamic viscosity with the relative sliding velocity per unit load or pressure exerted. The curve can be used to find the ideal lubricant speed for lubricated contacts that reduces the coefficient of friction. Different tribometres can be used for lubricant testing, with each one addressing a distinct lubrication regime [38].

2.4 Wear

Wear refers to the gradual material erosion or removal from a solid surface by another surface's mechanical action [39, 40]. Wear is primarily concerned with surface inter-actions, particularly material removal from a surface as a result of mechanical inter-action. The primary cause of this material removal or loss is the mechanical action between two sliding surfaces that are under the action of a load. Because wear is basically a universal phenomenon, it is rare for two solid bodies to touch or even slide over one another without some sort of quantifiable material loss or transfer [40]. One crucial difference that sets mechanical wear apart from other similar processes that produce comparable results is the requirement for mechanical action, specifically contact resulting from relative motion [41].

This wear definition does not include dimension loss as a result of plastic defor-mation, though wear has happened notwithstanding the absence of material removal. Additionally, impact wear—where there is no sliding motion; cavitation—where the

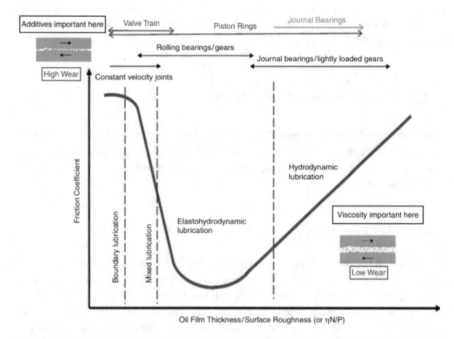

Fig. 2.3 Stribeck curve, indicating the different forms of lubrication regime. *Source* Wang and Chung [9]

counter body is a fluid; and corrosion—where the damage is as a result of chemical reaction rather than mechanical interaction, are not encompassed in this definition [42]. Wear of metallic materials and some other materials are caused by *plastic displacement of the material on and near the surface* and the *particles detachment* that makes up the wear debris, which has particles ranging in size from millimetres to nanometres [39, 41]. A number of factors, including temperature, rolling or sliding motion type, impact, static or dynamic loading types, and lubrication—particularly the deposition and wearing out of the boundary lubrication layer affect wear rate [2, 43].

2.4.1 Working Environmental Factors that Affect Wear

Loads (such as unidirectional sliding, reciprocating, rolling and impact loads), *temperature, speed, counter body type* and *type of contact* (such as lubricated or unlubricated contacts) are some of the factors in the working environment that can ascertain whether contact surfaces are protected against mechanical wear or not [2, 37]. Minimising friction and protecting lubricated contacts from mechanical wear is one of any lubricant's primary purposes. An important aspect of wear reduction is the capacity of the EP-type lubricants to avert scuffing, scoring and seizures under increasing applied loads [24, 44].

2.4.2   Classifications of Wear Modes

Based on the involved tribosystem, various classifications of wear modes (or types) can be encountered. The major *wear modes* can be categorised into four basic classifications [39, 40]. These basic classifications are abrasive, corrosive, adhesive and fatigue wear.

2.4.2.1   Abrasive Wear

Abrasive wear is a wear mode caused by hard, loose abrasive particles that are carried in by external impurities, particles implanted in one of the opposing surfaces, or wear particles generated due to adhesive wear that roll between two soft sliding surfaces [40]. Also, abrasive wear takes place when a softer surface glides over a harder, rougher one, creating grooves on the softer surface. Apart from their capacity to transfer particles to filtering systems that extract these particles from the circulating lubricating oil, oil characteristics in both scenarios have little direct impact on the quantity of abrasive wear that takes place. Abrasive wear can be avoided by either getting rid of the hard, rough component or making the intended surface that needs protection even harder, as it takes place when the abrasive materials are rougher and harder than the abraded surface [40].

2.4.2.2   Corrosive Wear

Corrosive wear, also known as chemical wear, is a wear mode that results from a synergistic chemical action with the rubbing or sliding action of metal surfaces that removes the corroded surface metal [45]. The wear that can happen to diesel engines' piston rings and cylinder walls when they burn high-sulphur fuels is a common illustration of corrosive wear. As the rings rub against the cylinder walls, the strong acids created by the burning of the sulphur can attack the metal surfaces, forming compounds that can be removed fairly easily. Rusting is a chemical reaction that takes place as a result of prolonged engine idleness, high moisture content, or the oil's inability to protect wet surfaces [24]. When equipment starts up, the rust is eliminated from contact areas, which leads to the loss of surface metal and the generation of particles that cause abrasive wear on the equipment.

2.4.2.3 Adhesive Wear

Adhesive wear is a wear mode that takes place when the lubricating film gets so thin that opposing surface asperities make contact with each other due to strong adhesive forces being produced at the interface of lubricated solid contacts [24, 40]. These adhesive forces might be as a result of load, speed, or temperature conditions [40]. Pressing two lubricated contact surfaces together results in intimate contact across several tiny areas or intersections, which are formed and broken continuously during sliding. When one of these intersections breaks away from its original interface, wear particles are created and finally break away. Adhesive wear, which is the most common wear mode, is detrimental because the mechanism's performance will deteriorate due to material loss and large wear particles may be formed in tightly fitted sliding contacts, potentially causing the early seizure of the mechanisms in their productive life. Adhesive wear, associated with sliding, is negligible, however, for sliding surfaces, adhesive wear typically begins adequately so fast that there is no room for fatigue wear to take place.

2.4.2.4 Fatigue Wear

Fatigue wear, also called surface-fatigue wear, is a wear mode caused by repeated high culminating stress attendant on the bounding surfaces that causes the formation of subsurface or fatigue cracks in either the moving or the unmoving surface [40]. As the propagation of the cracks progress, large particles detach from the surface and pitting starts, which leads to fatigue spalling or pitting, despite that the lubricating film maintains a complete surface separation and opposing asperities contacts are absent or comparatively negligible. Fatigue wear occurs in mechanical systems with a high degree of rolling like those of metal wheels on tracks, rolling element bearings and gears. The micropitting wear mechanism, which causes tiny pits in the surfaces of some gears and bearings, is a variation of fatigue pitting wear.

2.5 Summary

Friction is the force that resists the relative motion of interacting solid surfaces and material elements against each other. The microscopic mechanics involved in generating friction include adhesion, mechanical contacts of opposing surface asperities, deformation and or fracture of surface layers, and interference and local plastic deformation. The four basic friction types widely encountered are dry friction, fluid friction, mixed friction and boundary friction. Friction can be minimised between interacting surfaces by selecting appropriate materials with low adhesion properties, using lubricants and surface coatings, and subjecting the sliding interface to ultrasonic vibrations.

Lubrication is a technique used to interpose a lubricant between two surfaces to cushion the pressure generated by the opposing surfaces in close proximity to minimise friction and protect against the wear of one or both surfaces. Lubrication of mechanical systems such as bearings and gears or machine elements has several important functions, which include the formation of an elastohydrodynamic (EHD) or thin film for complete surface separation; reduction of friction and protection against the wear of interacting surfaces; transference of heat away from bearings, gears, or machine elements; vibration damping; prevention of the finished surfaces against corrosion; and acting in energy-transferring devices as shock-damping fluids.

The EHD, or thin lubricating film thickness, generated between the lubricated contacts is a function of the speed at which the surfaces interact, the load (or pressure exerted) by the opposing surfaces, the oil viscosity (or grease consistency), the operating temperature and contamination. In general, the formation EHD or thin-film requirement is the primary consideration in the selection of lubrication. For the formation of EHD, or thin lubricating films, to take place, the pressure of the lubricating film should be sufficiently high to trigger the interacting surfaces' elastic deformation.

The three basic varieties of lubrication are fluid-film lubrication—a lubrication type formed by interposing fluid film that completely separates sliding surfaces; boundary lubrication—a condition that exists between unlubricated and fluid-film lubrication sliding; and solid lubrication—a lubrication type in which a solid film is interposed between two surfaces in contact. Lubrication films are basically classified into three categories: fluid films, which are sufficiently thick films generated by the opposing motion of lubricated contact during normal operation that maintains complete surface separation; thin films, which are not sufficiently thick films formed during the opposing motion of lubricated contacts that maintain complete surface separation; and solid films, which are films formed by more or less permanently bonded solid lubricant to the bounding surfaces. The four basic lubrication regimes for self-pressure generating lubricated contacts are boundary, mixed, elastohydrodynamic and hydrodynamic lubrications.

Wear is the gradual material erosion or removal from a solid surface by another surface's mechanical action. The basic wear modes can be categorised into four classification: abrasive wear, which is a wear mode caused by hard, loose abrasive particles that are carried in by external impurities, particles implanted in one of the opposing surfaces, or wear particles generated due to adhesive wear that roll between two soft sliding surfaces; corrosive wear, which is a wear mode that results from a chemical action in combination with the rubbing or sliding action of metal surfaces that removes the corroded surface metal; adhesive wear, which is a wear mode that takes place when the lubricating film gets so thin that opposing asperities make contact; fatigue wear, which is a wear mode caused by repeated high culminating stress attendant on the bounding surfaces that causes subsurface or fatigue cracks to form in either the moving or the stationary surface.

References

1. Mate, C. M. (2002). On the Road to an atomic- and molecular-level understanding of friction. *MRS Bulletin, 27*, 967–971. https://doi.org/10.1557/mrs2002.303
2. Chattopadhyay, R. (2004). *Advanced thermally assisted surface engineering processes*. Kluwer Academic Publishers.
3. Hirani, H. (2016). *Fundamentals of engineering tribology with applications*. Cambridge University Press.
4. Runway Surface Friction (2010). Friction. *SKYbrary*. Retrieved from http://www.skybrary. aero/index.php/Runway_Surface_Friction
5. Bhushan, B., Israelachvili, J. N., & Landman, U. (1995). Nanotribology: Friction, wear and lubrication at the atomic scale. *Nature, 374*(6523), 607–616.
6. Hendrik, H., André, S., & Schwarz U. D. (2008). Principles of atomic friction: from sticking atoms to superlubric sliding. *Philosophical Transactions of the Royal Society A, 366*, 1383–1404. https://doi.org/10.1098/rsta.2007.2164
7. Menezes, P. L., Ingole, S. P., Nosonovsky, M., Kalias, S. V., & Lovell, M. R. (2013). T*ribology for scientists and engineers—From basics to advanced concepts*. Springer Science+Business Media.
8. Abdelbary, A., & Chang, L. (2023). *Principles of engineering tribology: Fundamentals and applications*. Academic Press. https://doi.org/10.1016/B978-0-323-99115-5.01001-X
9. Wang, O. J., & Chung, Y. (Eds.). (2013). *Encyclopedia of tribology*. Springer.
10. Lyashenko, I. A. (2011). Tribological properties of dry, fluid and boundary friction. *Technical Physics, 56*(5), 701–707.
11. Persson, B. N. J. (2000). *Sliding friction: Physical principles and applications*. Springer.
12. Taylor, R. I., & Sherrington, I. (2022). A simplified approach to the prediction of mixed and boundary friction. *Tribology International, 75*, 107836.
13. Anh, L. (2003). *Dynamics of mechanical systems with Coulomb friction*. Springer.
14. Bowden, F. P., & Tabor, D. (2001). *The friction and lubrication of solids*. Clarendon Press.
15. Friction Reduction (2023). Friction reduction. *Wikiwand*. Retrieved from https://www.wik iwand.com/en/Friction#Reduction
16. Bhushan, B. (2013). *Introduction to tribology*. Wiley.
17. Meng, T. (Ed.). (2014). *Encyclopedia of lubricants and lubrication*. Springer.
18. Britannica, T. (Ed.). (2023). Friction. *Encyclopedia Britannica*. Retrieved from https://www. britannica.com/science/friction
19. Kalpakjian, K., & Schmid, S. R. (2009). *Manufacturing engineering and technology* (6th Ed.). Pearson Prentice Hall.
20. Sivaprasad, P., & Davies, C. H. (2005). An assessment of the interface friction factor using the geometry of upset specimens. *Modelling and Simulation in Materials Science and Engineering, 13*(3), 355–360.
21. Friction Applications. (2023). Friction applications. *Wikiwand*. Retrieved from https://www. wikiwand.com/en/Friction#Applications
22. Lubrication. (2023). Lubrication. *Wikiwand*. Retrieved from https://www.wikiwand.com/en/ Lubrication#introduction
23. Britannica, T. (Ed). (2023). Lubrication. Encyclopedia Britannica. Retrieved from https://www. britannica.com/technology/lubrication
24. Pirro, D. M., Webster, M., & Daschner, E. (2016). *Lubrication fundamentals* (3rd Ed., Revised and Expanded). Routledge, Taylor & Francis Group.
25. Pirro, D. M., & Wessol, A. A. (2001). *Lubrication fundamentals* (2nd Ed. Revised and Expanded). Marcel Dekker, Inc.
26. Ajimotokan, H. A., & Mahamood, R. M. (Eds.). (2017). *Engineering workshop technology*. University of Ilorin Publishing House.
27. Harnoy, A. (2003). *Bearing design in machinery: Engineering tribology and lubrication*. Marcel Dekker Inc.

28. Ajimotokan, H. A. (2023). MEE 344: Tribology. Course guide, Department of Mechanical Engineering, University of Ilorin, Ilorin, Nigeria.
29. Srivastava, S. P. (2014). *Developments in lubricant technology*. Wiley.
30. Bruce, R. W. (2012). *Handbook of lubrication and tribology, Vol. II: Theory and Design* (2nd Ed.). CRC Press.
31. Miyoshi, K. (2001). *Solid lubrication: Fundamentals and applications*. Taylor & Francis Group.
32. Beall, C. J. (2002). Solid film lubricant. *Metal Finishing, 100*(1), 505–508. Retrieved from https://doi.org/10.1016/S0026-0576(02)82054-2
33. Machinery Lubrication. (2023). *When to use solid-film lubricants*. Retrieved from https://www.machinerylubrication.com/Read/29057/solid-film-lubricants
34. Speight, J. G. (2015). *Handbook of petroleum product analysis* (2nd ed.). Wiley.
35. Miyoshi, K. (1999). Considerations in vacuum tribology (adhesion, friction, wear, and solid lubrication in vacuum). *Tribology International, 32*, 605–616.
36. Sadeghi, F. (2010). Elastohydrodynamic lubrication. In T*ribology and dynamics of engine and powertrain: Fundamentals, applications and future trends*. Woodhead Publishing Ltd. https://doi.org/10.1533/9781845699932.1.171
37. Chauhan, A. (2016). *Non-circular journal bearings*. Springer.
38. Ogedengbe, T. S. (2018). *MEE 509: Tribology. Course guide*. Department of Mechanical Engineering, Elizade University, Ilara-Mokin, Nigeria.
39. Wear. (2023). Wear. *Wikiwand*. Retrieved from https://www.wikiwand.com/en/Wear
40. Britannica, T. (Ed.). (2023). Wear. *Encyclopedia Britannica*. Retrieved from https://www.britannica.com/science/wear
41. Akchurin, A., Bosman, R., Lugt, P. M., Drogen, M. (2016). Analysis of wear particles formed in boundary-lubricated sliding contacts. *Tribology Letters, 63* (2), 16. https://doi.org/10.1007/s11249-016-0701-z
42. Martin, P. M. (2011). *Introduction to surface engineering and functionally engineered materials*. Scrivener Publishing LLC.
43. Popov, V. L. (2018). Is tribology approaching its golden age? Grand challenges in engineering education and tribological research. *Frontier in Mechanical Engineering 4*, 16. https://doi.org/10.3389/fmech.2018.00016
44. Blau, P. J. (2017). *Tribosystem analysis: A practical approach to the diagnosis of wear problems*. CRC Press, Taylor & Francis Group.
45. Li, D. Y. (2013). Corrosive wear. In Q. J. Wang, Y. W. Chung (Eds.), *Encyclopedia of tribology*. Springer. https://doi.org/10.1007/978-0-387-92897-5_866

Chapter 3
Viscosity of Fluid Lubricants

Abstract The objectives of this chapter are to: (i) Define the terms viscosity, dynamic viscosity and kinematic viscosity; (ii) Highlight the fundamental principles of viscosity in the selection of a suitable oil for a given application with respect to temperature; (iii) Identify and describe the Newton's law of viscosity; (iv) Identity, outline and discuss the classifications of engine oil viscosity according to their viscosity ranges and grades; (v) Outline and discuss manual transmissions and axles viscosity classifications according to their viscosity ranges and grades; (vi) Define the term viscosity index; and (vii) Outline the fundamental principles of viscosity index with respect to the rate of change in viscosity relative to temperature fluctuations, among others.

Keywords Dynamic viscosity · Kinematic viscosity · Newton's law of viscosity · Engine oil · Multigrade oil · Monograde oil · Viscosity index

3.1 Introduction

The lubricating oil's viscosity is probably its most significant singular property because it influences the *formation of elastohydrodynamic* or *thin films* between lubricated contacts of rolling elements and raceways or sliding interfaces, respectively, under both thin and fluid-film conditions; governs the *extent of frictional effect that would be experienced* between interacting surfaces; influences *heat generation* in gears, bearings and machine elements; controls the *lubricating oil's sealing effects*; governs the *lubricating oil's consumption rate or loss*; and ascertains the *ease with which an engine at cold ambient conditions can be cranked* [1–5]. The use of lubricating oil with the appropriate viscosity is one of the basic requirements for successful operation of any mechanical piece of equipment.

© The Author(s), under exclusive license to Springer Nature Switzerland AG 2024

H. A. Ajimotokan, *Principles and Applications of Tribology*,
Manufacturing and Surface Engineering, https://doi.org/10.1007/978-3-031-57409-2_3

3.2   Viscosity

Viscosity is defined as the property of a fluid that causes it to resist flowing under load. It can also be expressed as the property of a fluid that determines its shearing stress resistance. The viscosity of any fluid causes resistance to fluid flow (under load or stress) predominantly because of *cohesion and molecular momentum exchange* among fluid layers, and as fluid flow takes place, these resistive effects to fluid flow materialise as shearing stresses between the fluid's moving layer [6]. The viscosity of any fluid is sensitive to small *temperature changes* [4, 5, 7]. With liquids, viscosity *decreases with an increasing temperature* because the shear stress owing to inter-molecular cohesion diminishes with an increasing temperature. For gases, their inter-molecular cohesion is insignificant, and the shear stress can be because of the molecular momentum exchange between the gaseous fluid layers, normal to the direction of motion [8, 9]. As these molecular activities increase with an increase in temperature, the viscosity of the gas increases proportionately [9]. Unlike with temperature, all fluids' viscosities never appreciably *change with pressure* under ordinary conditions. Conversely, some lubricating oils' viscosities have been observed to increase with an increasing pressure [9].

   With an increase in temperature, *mineral* (or petroleum) and *synthetic lubricating oils'* viscosities appreciably diminish (i.e., they become thinner), but when the lubricating oils are then cooled to their original temperature, the higher viscosity is restored [4, 7, 9]. The synthetic oil's viscosity is comparatively less sensitive to variations in temperature (compared with mineral oils). With an increasing temperature, the viscosity of synthetic oil also diminishes [10]. Viscosity is the *primary consideration* to take into account when choosing a suitable lubricating oil for a given application. For both thin- and fluid-film circumstances, the lubricating oil's viscosity should be *sufficiently high to generate appropriate lubricating films* between interfacing surfaces, but not *too high to cause excessive frictional losses*. Also, it is essential to take into consideration the oil's apparent operational temperature in an engine because lubricating oils' viscosity changes with temperature. Other factors to be taken into account include whether or not an engine would be cranked under low ambient temperatures. Every fluid lubricant's viscosity might be either dynamic or kinematic [5, 9].

3.2.1   Dynamic Viscosity

Dynamic viscosity can simply be expressed as the viscosity that relates the shear stress in a fluid to its shear rate; that is, if the lubricating oil's viscous shear stress is proportional to its shear rate, the factor of proportionality is termed '*dynamic viscosity*'. Mathematically, dynamic viscosity can be expressed using Eq. 3.1:

$$\tau = \eta \frac{du}{dz}, \tag{3.1}$$

where η denotes the dynamic viscosity, τ is the shear stress and $\frac{du}{dz}$ is the shear rate. Hence, at the same shear rate, relative thicker oils experience larger shear pressures due to their higher viscosity values. Typically, any fluid's dynamic viscosity can be estimated at high shear conditions, employing a cone on a plate or the cylinder viscometre that measures the viscous shear torque between two cylinders. Through certain well-established functions of interpolation from those of Reynolds or Vogel and Cameron, the viscosity can be calculated using in-between temperatures for known viscosity at two reference temperatures. Kinematic viscosity is typically used to characterise lubricants because it is more suitable to measure viscosity in a way that the density of the lubricant influences the measurement. Dynamic viscosity, in SI unit, is measured in *Pascal-seconds* (Pa s) but commonly reported in *poise* (*P*) or *centipoise* (cP) (with 1 cP = 0.01 *P* = 0.001 Pa s). In calculations concerning bearing designs and oil flow, the quantity used most frequently is dynamic viscosity, which only depends on the fluid's internal friction [5].

Figure 3.1 depicts the basic concept of dynamic viscosity, which indicates a plate being pulled at a constant speed over a stationary surface on a film of lubricant. Figure 3.1a depicts a pictorial illustration of how the oil sticks to both stationary and moving surfaces. The oil film that touches the moving surface moved at the same velocity U as the moving surface, whereas the oil that touches the stationary surface remained at zero velocity as the simplified view depicted (see Fig. 3.1b). The oil film can be seen to consist of several layers in between, each of which is drawn out separately by the layer above it at a velocity that is only a small amount of the velocity U that is proportional to their height above the stationary plate. To reduce the fluid layers' frictional effects between them, a force F is exerted to propel the moving plate. Since fluid's friction is brought about by viscosity, the lubricating oil's viscosity is inversely proportional to the force. Thus, the viscosity of a substance might be determined from the measurement of the force required in displacing the fluid friction in oil films with known dimensions. The viscosity ascertained using this approach is termed '*dynamic viscosity*', also known as '*absolute viscosity*' [4].

3.2.2 Kinematic Viscosity

Kinematic viscosity can be defined as the quotient of a fluid's dynamic viscosity and its density, measured in consistent units at the same temperature [11]. Mathematically, kinematic viscosity can be expressed using Eq. 3.2:

$$\nu = \frac{\eta}{\rho}, \tag{3.2}$$

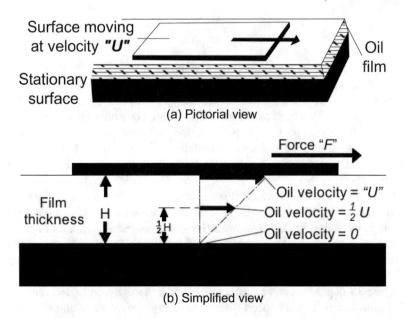

Fig. 3.1 Concept of dynamic viscosity **a** pictorial view and **b** simplified view. *Source* Pirro and Wessol [4]

where v denotes the kinematic viscosity and ρ is the density of the fluid. Its SI unit is *square millimetres per second* (mm²/s), but mostly reported in *Stokes* (St) or *centistokes* (cSt) (with 1 cSt = 0.01 St = 1 mm²/s). Dynamic viscosity, in cP, may be converted to kinematic viscosity, in *cS*, by dividing the dynamic viscosity by the fluid's density, in *gram per cubic centimetres* (g/cm³), at an equal temperature. Kinematic viscosity, in cS, may be converted to dynamic viscosity, in cP, by multiplying the kinematic viscosity by the fluid's density, in g/cm³, at an equal temperature. Kinematic viscosities, in mm²/s, may be converted to dynamic viscosities, in Pa s, by multiplying the kinematic viscosity by the fluid's density, in g/cm³ and dividing the product by 1000. The physical principle of measurement depends upon the fluid flow rate under gravity through a capillary tube. When the viscosities of a fluid are known at two reference temperatures, the viscosities at intermediate temperatures may be determined employing *Ubbelohde-Walther's interpolation function,* an established standard adopted by the American Society for Testing and Materials, ASTM D341 [5, 12].

3.3   Newton's Law of Viscosity

Newton's law of viscosity states that the shear stress on a fluid's element layer is directly proportional to its shear strain rate. Mathematically, the law can be expressed using Eq. 3.3 [6]:

$$\tau = \mu \frac{\mathrm{d}u}{\mathrm{d}y}, \tag{3.3}$$

where τ denotes the shear stress, μ is the proportionality constant called the coefficient of viscosity and $\frac{\mathrm{d}u}{\mathrm{d}y}$ is the rate of shear strain.

3.4   Classifications of Lubricants' Viscosity

There are three different viscosity-numbering systems used to categorise viscosity of fluid lubricants based on ranges of viscosity; two of these systems and the third are used for automotive lubricants and industrial oils, respectively [4]. Lubricants used in industrial machinery and internal combustion engines have been categorised over time based on different viscosity grades [6]. The American Gear Manufacturers' Association (AGMA) grades for gear oils and the Society of Automotive Engineers (SAE) grades for crankcase (also known as engine) and gear oils are two examples of these classifications of viscosity grades. Two temperature measurements (i.e., Celsius and Fahrenheit) may be used in practically all of them, which adds to the complexities.

Although all classes of viscosity grades can, to some extent or another, serve significant objectives, practically all lubricant users choose the SAE grades and use their methodology as a basis in product selection. Since the SAE establishes the tangible viscosity grade of lubricants, it is particularly widely employed. In accordance with the SAE Standard J 300, engine oils are categorised according to their viscosities at low shear rates and high temperatures ($100°C$), high shear rates and high temperatures ($150°C$), and both low and high shear rates at low temperatures ($-5°C$ to $-50°C$) [13]. For calculating the proper engine oil viscosities as well as for identifying the internal combustion engine oils' viscosities, engine manufacturers frequently use these SAE systems.

3.4.1   Engine Oil Viscosity Classifications

There are two *basic classifications of engine oil viscosity grades.* These classifications are *multigrade* and *monograde oil* [4]. Examples of multigrade oil, also called multiviscosity oil are SAE 5W-30, SAE 15W-40, SAE 20W-50, etc., and monograde oil is SAE 30, SAE 40, etc. In the multigrade classifications, SAE grade with the *suffix letter W*, which denotes winter, is predominantly intended for *usage under low ambient temperatures*, whereas in the monograde classifications, SAE grade that has *no suffix letter W* is formulated for usage where *low ambient temperatures* would never be experienced [8]. The '*15W*' in the SAE 15W-40 denotes the low-temperature limits of the 15W grade while the '*40*' denotes the high-temperature limits of the 40 grade.

The multigrade oil might be employed for a *year-round service* in automobile engines, excluding certain two-stroke engines of the diesel types, due to its oil formulation, which meets the engine low-temperature limits (such as –5°C to – 50°C limits) of the 5W or 15W grades and high-temperature limits (such as 100°C or 150°C limits) of the 30 grade or 40 grade, respectively [14]. Such an oil formulation is designated SAE 5W-30 grade or SAE 15W-40 grade, respectively, and is called *multigrade oil* [3, 14]. In general, multigrade oil types need the addition of viscosity index (VI) improvers to the mineral or synthetic-based lubricating oil base stocks.

At comparable shear stresses observed in engine bearings, the high-temperature-high-shear rate viscosity can be determined at 150°C based on ASTM D 4683 or D 4741 standard method [4, 15, 16]. The high-temperature-high-shear rate number denotes the short-term shear stability of the VIs employed for the formulation of the multigrade engine oils [13]. According to the SAE J300 categorisation, oil of SAE 40 grade classification should possess a high-temperature-high-shear rate viscosity greater than 3.69 cP at 150°C to function as a 40 grade oil at engine running circumstances [13]. ASTM D 5293 standard—a multitemperature simulation method under cold cranking conditions, is used to assess the high-shear rate viscosity at low temperatures. ASTM standard D 4684 is used to determine low-shear-rate viscosity at low temperatures. Both of their tests' results have been depicted to be consistent with engine cranking and the engine oil's capacity to maintain pressure at low temperatures [4].

3.4.2 Manual Transmissions and Axles Lubricant Viscosity Classifications

According to the recommended practice SAE J306, lubricants can be classified for application in *automotive manual transmissions and drive axles* based on their viscosity at 1000°C (2120°F) and at the highest temperature that they attain 150,000 cP (150 Pa s) [4]. When cooled, they can be determined using the Brookfield viscometre in accordance with standard test method ASTM D 2983 for apparent viscosity at low temperature [4, 5]. The manual transmissions and axle lubricant viscosity categories and their limits are listed in Table 3.1. Under this system, formulation of multigrade oils like SAE 80W-90 and SAE 85W-140 can be conducted. Based on the test data of pinion bearings' lubrication failures for a particular axle design, which depicted that the lubrication failures of pinion bearings occurred when

Table 3.1 Manual transmissions and axles lubricant viscosity classifications

SAE viscosity grade	Maximum temperature for viscosity of 150,000 cP (°C)	Viscosity at 100 °C (cSt)	
		Minimum	Maximum
70W	– 55	4.1	–
75W	– 40	4.1	–
80W	– 26	7.0	–
85W	– 12	11.0	–
90	–	13.5	24
140	–	24.0	41
250	–	41.0	–

Source Pirro and Wessol [4]

the lubricant viscosity exceeded a limitation viscosity of 150,000 cP [4]. Because other axle designs and transmissions can exhibit limitation viscosities that are greater or lower, manufacturer of gears must be obliged to identify the precise grades that would perform well in various environmental situations.

3.5 Viscosity Index

V*iscosity index* is an arbitrary, unit-less scale employed for measuring changes in lubricating oil's viscosity relative to temperature fluctuations [4, 6]. In other words, the VI is a technique for assigning a numerical value to the change in fluid's viscosity rate relative to temperature fluctuations, on the basis of comparing the comparative viscosities' rates of change for two randomly chosen oil types, which are widely different in their viscosity characteristics. A high VI and a low VI suggest a comparatively *low rate of viscosity change* with temperature and a comparatively *high rate of viscosity change* with temperature, respectively, for instance, considering two different oils of high and low VI at room temperature having the same viscosity. With an increasing temperature or at higher temperatures under load, the oil of high VI thins out relatively less (i.e., resist flowing relatively more) and, thus, exhibits a higher viscosity compared with the oil of low VI. Oils' VIs can be computed from viscosities estimated at two temperatures using established ASTM table [6]. These tables are available on the basis of estimated viscosities at both 100°F and 212°F, as well as at both 40°C and 100°C. Typically, VI is utilised to characterise the lubricating oils' *viscosity-temperature behaviour*.

Various lubricating oil types have diverse rates of viscosity change with temperatures, e.g., distillate oil from naphthenic base crude exhibits a greater rate of viscosity change with temperature compared with distillate oil from paraffin crude [4]. Finished petroleum-based lubricating oils refined by employing traditional techniques have varied VI from slightly less than zero to somewhat over a 100. But petroleum-based oil base stocks made by exceptional hydro-processing methods might possess VIs over a 100 [7]. Certain synthetic-based oils possess VIs at both lower and over this range. Additives—organic or inorganic compounds that are suspended as solids or dissolved in finish lubricating oils to impart certain characteristics to the lubrication oils, known as VI improvers might be blended into lubricating oils for enhancing the VIs. The VI improver is a long chain, high molecular weight polymer that functions by causing the lubricating oil's comparative viscosity to rise more under high temperatures compared to under low temperatures.

Though, the VI improver is certainly not stable on all occasions in lubricating environments when exposed to thermal or shear stresses. Consequently, this additive should be employed with necessary caution to ensure acceptable viscosity during the expected interval of service for the purpose which it is intended. In numerous service types with more or less constant operating temperature, the oil's VI is of little or no concern. Although, in services where the oil's operating temperature might range over a broad limit, like in passenger automobile engines, it is perhaps safe to recommend that the oil's VI employed must be as high as feasible and compatible with other performance metrics.

3.6   Summary

Viscosity of lubricating oil is the most significant singular property because it influences the formation of lubricating films; determines the amount of friction that would be experienced; influences heat generation in bearings, cylinders and gears; controls the lubricating oil's sealing effects; governs the lubricating oil's consumption rate or loss; and determines the ease with which an engine can be cranked.

Viscosity is the fluid's property that causes it to resist flowing under load. The dynamic viscosity is the viscosity that relates the shear stress in a fluid to its shear rate; that is, if the lubricating oil's viscous shear stress is proportional to its shear rate, the factor of proportionality is the dynamic viscosity. The kinematic viscosity is the quotient of a fluid's dynamic viscosity and its density, measured in consistent units at the same temperature.

The two classifications of engine oil viscosity grades are multigrade and monograde oils. The viscosity index is a unitless, arbitrary scale employed for measuring changes in lubricating oil's viscosity relative to fluctuations in temperature.

References

1. Ahmed, N. S. and Nassar, A. M. (2011). Lubricating oil additives. In C. Kuo (Ed.) *Tribology: Lubricants and lubrication.* InTech Europe. Retrieved from http://www.intechopen.com/books/tribology-lubricants-and-lubrication/lubricating-oil-additives

2. Rizvi, S. Q. A. (2009). *A comprehensive review of lubricant chemistry, technology, selection and design.* ASTM International.

3. Margareth, J. S., Peter, R. S., Carlos, R. P. B., & José, R. S. (2010). Lubricant viscosity and viscosity improver additive effects on diesel fuel economy. *Tribology International, 43,* 2298–2302.

4. Pirro, D. M., & Wessol, A. A. (2001). *Lubrication fundamentals* (2nd Ed., Revised and Expanded). Marcel Dekker, Inc.

5. Pirro, D. M., Webster, M., & Daschner, E. (2016). *Lubrication fundamentals* (3rd Ed., Revised and Expanded). Routledge.

6. Wang, O. J., & Chung, Y. (Eds.). (2013). *Encyclopedia of tribology.* Springer.

7. Bruce R. W. (2012). Handbook of lubrication and tribology, Vol. II: Theory and design (2nd Ed.). CRC Press.

8. Rajput, R. K. (1998). *Fluid mechanics and hydraulic machines.* S. Chand & Company Ltd.

9. Ajimotokan, H. A. (2023). MEE 344: tribology. Course Guide, Department of Mechanical Engineering, University of Ilorin, Ilorin, Nigeria.

10. Speight, J., & Exall, D. I. (2014). *Refining used lubricating oils* (1st ed.). CRC Press.

11. Menezes, P. L., Ingole, S. P., Nosonovsky, M., Kalias, S. V., & Lovell, M. R. (2013). Tribology for scientists and engineers—From basics to advanced concepts. Springer Science+Business Media.

12. Doll, G. L. (2023). Rolling bearing tribology: tribology and failure modes of rolling element bearings. Elsevier. Retrieved from https://doi.org/10.1016/B978-0-12-822141-9.09993-3

13. Mortier, R. M., Fox, M. F., & Orszulik, S. T. (2010). *Chemistry and technology of lubricants.* Springer.

14. Pillon, L. Z. (2010). *Surface activity of petroleum derived lubricants* (1st ed.). CRC Press.

15. Neveu, C. D., Sondjaja, R., Stöhr, T., Schimossek, K., & Iroff, N. J. (2016). Lubricant and fuel additives based on polyalkylmethacrylates. In *Reference module in materials science and materials engineering.* Elsevier. Retrieved from https://doi.org/10.1016/B978-0-12-803581-8.01544-7

16. Meng, T., & Dresel, W. (2007). *Lubricants and lubrication.* Wiley-VCH.

Chapter 4
Surface Studies and Functionality Characterisation

Abstract The objectives of this chapter are to: (i) Define the term surface topography; (ii) Outline and highlight the classifications of surface irregularities; (iii) Identify, outline and discuss the components of surface topography; (iv) Identify, outline and describe the different methods for surface topography characterisation; and (v) Define the term surface asperity.

Keywords Topography and qualities · Surface topography · Surface irregularity · Surface topography characterisation · Surface asperities

4.1 Introduction

A surface establishes an object's exterior boundaries and interacts in a variety of ways within its immediate environment. Whether a material is functionally engineered or natural, its *surface topography* plays a critical role in determining its properties [1]. The topography is one of the key physical characteristics of surfaces that influence significantly the material's technical and bulk properties [2]. Surface imperfections are frequently on the nanoscale, but macroscopic measurements can still be employed to observe their impact. Numerous fields, including materials science and engineering, tribology and engine condition monitoring, among others, have given increased importance to the characterisation of surface topography. Besides, the by-products of the surface topography characterisation result in *solid surface studies and surface functionality characterisation*, providing more information on surface functionalities [3]. These data on surface functionalities can include, among other things, information on friction, lubrication, wear, fatigue strength, joint tightness, heat and electrical current conductance, cleanliness, reflectivity, sealing, positional accuracy, load-carrying capacity, corrosion resistance, and paint and coating adhesion [3]. Therefore, the implementation of tribological interaction depends heavily on knowing the nature of the interacting contacts between rubbing or sliding surfaces and their functionalities [4].

© The Author(s), under exclusive license to Springer Nature Switzerland AG 2024   39
H. A. Ajimotokan, *Principles and Applications of Tribology*,
Manufacturing and Surface Engineering, https://doi.org/10.1007/978-3-031-57409-2_4

4.2   Surface Topography and Qualities

Surface topography, also called surface texture, is defined as a surface's local deviation from a perfectly flat plane [5, 6]. During sliding, surface topography is a vital factor that *controls friction and transfer layer formation*, which might be isotropic or anisotropic [6]. Occasionally, based on the topography of a surface, stick–slip friction phenomena might be experienced during sliding [7]. Every manufacturing operation creates a surface topography, which is influenced by a number of factors. In general, as the surface topography becomes smoother, the cost of producing a smooth surface rises. Because the surface topography characteristics are abstract in nature, a material with a range of surface roughness is generally manufactured employing various manufacturing operations. Typically, the operation is often optimised to make sure that the consequent topography is usable.

4.2.1   Surface Topography Characteristics

Surface topography characteristic, simply called surface irregularities, is largely classified into two basic groups [3]. These basic groups are *micro-irregularities* like roughness and waviness, and *macro-irregularities* like form errors, position errors and size deviation. Figure 4.1 depicts the classifications of surface irregularities. The *surface qualities* or *features*, such as roughness, waviness and microcracks have substantial influence on the final product performance [3]. At large, the surface qualities might be delineated as profiles, areas, forms and volumes.

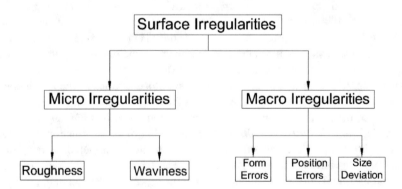

Fig. 4.1 Classifications of surface irregularities. *Source* Wang and Chung [3]

4.2.2 Components of Surface Topography

In the world of surface finishes, the *form surface quality* is defined by the long-wavelength components and the *surface roughness* is defined by the short-wavelength components [3]. There are four predominant components of surface topography that result from the usual manufacturing or machining processes of a workpiece. These are *lay*, which refers to the prevalent surface pattern's direction. Generally, the lay is ascertained by the employed manufacturing or machining processes; *roughness*, which refers to the closely spaced irregularities on a surface that are in-built during the manufacturing process after exempting the components of waviness and form. Typically, roughness is determined by the actual machining agent, such as cutting tool marks, grinding wheel grits and welding sparks; *waviness*, which is the feature component of a surface quality in which the roughness is overlaid or superimposed. The more widely spaced irregularities that make up waviness are typically caused by warping, machine or workpiece vibrations, deflection or heat treatment, numerous sources of material strain and extraneous influences; and *form*, which is the typical shape of the surface that depicts how the variations in surface quality can vary depending on the degree of roughness and waviness. *Form errors* are typically used to describe deviations from the ideal form [3].

In tribology, the key component among other components of the surface irregularities is the surface roughness due to the information it can provide on lubrication, wear, component life-time prediction and so on.

4.2.3 Tools for Surface Topography Characterisation

Surface topography can be characterised using two basic approaches. These approaches include both *contact* and *non-contact methods* [3, 8]. Besides, it is worth of note that to characterise surface topography, the computational method is also a valuable supplementary tool. To deal with the multiscale nature of the surface quality classification, computational methods are employed for characterising surface topography in conjunction with the use of other tools.

4.2.3.1 Contact Methods

To characterise the surface topography of any class of specimen, the *contact method* uses a measuring instrument or measurement stylus that is positioned in direct contact with the specimen and dragged across its surface [3]. The contact methods include tools like tactile profilometry and scanning probe microscopy. A *tactile profilometre* has the ability to characterise surface topography in two- and three-dimensions on a microscopic scale. The fundamental idea behind it is to scan a tip across the specimen's surface while applying pressure and directly determining the surface-height

information from the variation in tip height. As a result, the height worth in relation to every lateral sampling point reveals the roughness profile. The limitations
imposed by scanning tip geometries are the tactile profilometre's drawbacks, though.
Additionally, the tip's pressing force damages or leaves scan tracks on delicate
surfaces.

The tactile profilometres, however, are unable to measure down to nanometre scale
level. To examine the minute details at *the atomic and nanometre scales*, scanning
probe microscopes are used, such as atomic force and scanning tunnelling microscopes [3]. The scanning tunnelling microscope, like the tactile profilometre, scans
across the surface using a metal tip with a very small gap above the specimen's
surface, about 0.2 μm, which is the specified range of the detached distance. A
tunnelling current is induced and passes through the gap when voltage is supplied
between the two electrodes—the metal tip and an electrically conducting specimen.
To maintain the scanning tip at a specific height level throughout scanning, the
tunnelling current is adjusted. As a result, the surface topography influences how the
tunnelling current changes. A constant tunnelling current can be kept when a feedback
control system is used with the scanning tunnelling microscope. The surface-height
topography is then characterised based on the tip's permissible movements.

The *atomic force microscope* can be employed to characterise a variety of different
types of surfaces, in contrast to the scanning tunnelling microscope, which is limited
to only electrically conductive specimens [9]. There are three common modes of the
atomic force microscopy. These are *contact, non-contact* and *intermittent contact
modes* [9]. In the contact mode, the specimen is scanned over by the atomic force
microscope's sharp tip, which has a radius of about 20–50 nm and is mounted on
one end of a low-strength cantilever [10]. In this mode, the cantilever experiences
a constant contact or normal force of 1 nN, which causes the cantilever to deflect
in relation to the topography of the surface. The cantilever's spring constant and
its deflection's variation are then used to calculate the surface height. The applied
voltage in the non-contact mode is maintained (i.e., kept constant) at a detached
distance of roughly 1 nm between the specimen and the atomic force microscope's
sharp tip. As a result, the cantilever's deflection variation produces the surface-height
topography.

The low contact force and high stiffness of the cantilever, however, may be the
cause of the weak signal observed in the atomic force microscope's non-contact
mode. The *intermittent contact mode* of atomic force microscopy has been developed for comparative advantages in both the contact and non-contact modes. In the
intermittent contact mode, the cantilever of the atomic force microscope is oscil-
lated to assist in the decline of the lateral force, which is predominantly responsible
for the damage mostly done by the scanning tip. As a result, measurements can be
established in a less destructive and more effective manner. The atomic force microscope may also be employed to determine additional forces-related parameters, like
electrostatic attraction, chemical affinity, friction, lubrication, wear and adhesion.

4.2.3.2 Non-contact Methods

The *non-contact method*, also called non-destructive method, is a non-contact approach based on optical techniques [3]. When delicate surfaces need to be examined, non-contact methods can fulfil this very important requirement. These non-contact methods comprise, among others, interferometry, electron microscopy, focus variation and photogrammetry. Using a diamond stylus profilometre, that runs perpendicular to the surface, is the most popular non-contact method. Without making direct physical contact with the specimen's surface, optical-based techniques like confocal microscopy, white-light interferometry and electron microscopy can characterise surface topography. The application of optical-based techniques is on the rise because of the speed and capability of conducting non-destructive measurements over *large measurement ranges with high resolutions.* Optical microscopes, such as reflection confocal microscopes, can measure surface roughness owing to laser-speckle effects by using an optional light source called a white light-emitting diode (LED). The focus-detection method is the main method used in confocal microscopy, and the confocal microscope is unlike the conventional microscope in which the zone being characterised is minimised, one at a time, to just a focal point. The focal point of a surface must be scanned to characterise the whole surface.

The Nipkow disk rotation is aided in the lateral plane by the creation of a raster scan of multifocal points at real-time rate measurement. A piezoelectric transducer is used to accomplish the scanning in the axial direction, also known as the z or height direction. Primarily, the focal point of a confocal microscope functions similarly like the sharp tip of a scanning probe microscope. The objective's numerical aperture, which optically determines the lateral resolution, affects the diffraction limit's focal radius. Large amounts of lateral plane data in the vertical direction are accumulated through the z-direction scan and include details on the height of the surface topography relative to the coordinates of the lateral sampling (x, y). The distribution of output light intensity that corresponds to the maximum peak intensity is ultimately used to estimate the surface's height.

A white-light interferometer is appropriate for measurements of the irregularity at high resolution on smooth surfaces [3]. Its measurement method depends on producing fringe interferograms, which are made up of both dark and bright bands. At least two light beams, one from the specimen and the other from the white-light interferometer's reference mirror, interfere to produce the fringes. When the surface is in focus and the fringe contrast is at its maximum, the interference pattern is created as a function of the path length difference between the two light beams. For instance, when two beams follow paths that are precisely the similar length, they would impinge on one another in a constructive manner. The comparative tilt angle between the reference mirror and the specimen determines the number of fringes and their spacing. A charge-coupled device records a large intensity output per frame through scanning in the z-direction. The height data at every lateral sampling point on the specimen's surface is then ascertained through data analysis done using an interferometric phase mapping programme.

4.3   Surface Asperity

Surface asperity, simply called asperity can be described as the small projections, unevenness or roughness of a surface [7, 11]. Even surfaces polished to a mirror finish are not truly smooth at the atomic scale because they have asperities that are rough, sharp, or rugged projections [11]. Surface asperities can be found at different scales, frequently in fractal or self-affine geometry and the structures' fractal dimension has been linked to the friction and contact mechanics present at surface interfaces [12].

4.4   Summary

Surface topography is the surface's local deviations from a perfectly flat plane. Surface topography is largely classified into micro-irregularities like roughness and waviness, and macro-irregularities like form errors, position errors and size deviation.

The four predominant components of surface topography that results from usual manufacturing processes of a workpiece are lay, which refers to the direction of the predominant surface pattern, generally ascertained by the employed manufacturing operation; roughness, which refers to the closely spaced irregularities on a surface that is inherent in the manufacturing process; waviness, which is the feature component of a surface quality in which the roughness is overlaid or superimposed and form, which refers to the typical shape of the surface, depicting the variations of a surface quality as a function of roughness and waviness.

Both contact methods, which involve dragging a measurement stylus (profilometre) across the surface to determine surface topography, and non-contact methods, which are based on optical techniques and enable surface topography characterisation without making physical surface contact with the specimen, can be used to characterise surface topography.

Surface asperity is the small projections, unevenness or roughness of a surface.

References

1. Martin, P. M. (2011). *Introduction to surface engineering and functionally engineered materials*. Scrivener Publishing LLC.
2. Stout, K. J. (1994). *Three dimensional surface topography: Measurement interpretation and applications*. Penton Press.
3. Wang, O. J., & Chung, Y. (Eds.). (2013). *Encyclopedia of tribology*. Springer.
4. Meng, T. (Ed.). (2014). *Encyclopedia of lubricants and lubrication*. Springer.
5. Gopal, L. and Sudarshan, T. (2023). Functional surfaces through texture management. *Surface Engineering, 39*(3), 239–244. https://doi.org/10.1080/02670844.2023.2225004
6. Leach, R. (Ed.). (2011). *Optical measurement of surface topography*. Springer.
7. Menezes, P. L., Ingole, S. P., Nosonovsky, M., Kalias, S. V., & Lovell, M. R. (2013). *Tribology for scientists and engineers—From basics to advanced concepts*. Springer Science+Business Media.

8. Abdelbary, A., & Chang, L. (2023). *Principles of engineering tribology: Fundamentals and applications.* Academic Press. https://doi.org/10.1016/B978-0-323-99115-5.01001-X

9. Voigtlander, B. (2019). *Atomic force microscropy.* Springer Nature.

10. Bhushan, B., Fuchs, H., & Tomitori, M. (Eds.). (2008). *Applied scanning probe methods IX— Characterization.* Springer.

11. Ajimotokan, H. A. (2023). MEE 344: Tribology. Course Guide, Department of Mechanical Engineering, University of Ilorin, Ilorin, Nigeria.

12. Hanoar, D. A., Gan, Y., & Einav, I. (2015). Contact mechanics of fractal surfaces by spline assisted discretisation. *International Journal of Solids and Structures, 59,* 121–131. https://doi.org/10.1016/j.ijsolstr.2015.01.021

Chapter 5
Lubricants and Materials for Tribological Applications

Abstract The objectives of this chapter are to: (i) Define the term lubricants and identify, outline and discuss the principal types of lubricants; (ii) Identify, outline and describe the three typical inorganic compound kinds that could be employed as solid lubricants; (iii) Identify, outline and discuss those significant properties of commercial fluid lubricants; (iv) Identify, outline and discuss the reasons the effectiveness of the lubricating oil deteriorates over time or requires replacement after some time of usage; (v) Outline and discuss the lubricating oil selection and materials for tribological applications; (vi) Define the term additives and outline their roles and functions within lubricants; (vii) Identify, outline and describe the commonly used additives; (viii) Outline and discuss the tribology of rolling elements and applications; and (ix) Analyse, derive and discuss the power absorbed to overcome the viscous resistance due to the lubricating oil's viscosity in rolling bearings, such as journal, foot-step and collar bearings.

Keywords Liquid-oily lubricants · Solid lubricants · Gaseous lubricants · Lubricating oil deterioration · Lubricant selection · Additives · Rolling element tribology

5.1 Introduction

Mechanical systems like bearings and gears and numerous machine elements, among others are lubricated using a variety of grease, oils and exceptionally, in rare instances, solids and gaseous lubricants [1]. Grease is used to lubricate the majority of rolling element bearings, gears and machine elements because it *offers efficient lubrication, operates with very simple sealing* and *does not need elaborate supply systems* [1]. Bearings, gears and machine elements typically work well with just a very thin film of lubricating grease or oil [1–3]. However, oil lubrication is essential for use in high-speed mechanical systems, such as gears and internal combustion engines, to form adequate films or remove heat from their bearings [3, 4]. Understanding the

H. A. Ajimotokan, *Principles and Applications of Tribology*,
Manufacturing and Surface Engineering, https://doi.org/10.1007/978-3-031-57409-2_5

physicochemical and tribological interactions between the lubricant and the lubricated contacts (or materials) involved in rubbing or sliding interfaces is essential for giving the rolling element bearings, gears and machine elements an acceptable, smooth-running, satisfactory life.

5.2 Lubricants

A *lubricant* refers to any class of substance that can be liquid, solid, or exceptionally gaseous films interposed between interfacing surfaces in relative motion to minimise friction and protect against wear by providing an acceptable, smooth-running, satisfactory life for lubricated contacts [5, 6]. Primarily, lubricants *control friction and protect against wear*, but could and generally do carry out several other functions that differ with applications, which are mostly related [6]. The predominant objective of using lubricants is to *maintain suitable films* that are adequate in *minimising friction* and *protecting against wear* between lubricated contacts by providing *elastohydrodynamic* (EHD) or *thin films* between rolling elements and raceways, or sliding interfaces, respectively, under both thin- and fluid-film conditions [1, 2]. These EHD films formed on rubbing surfaces or thin films on sliding surfaces should be sufficiently thick to minimise friction and maintain complete surface separation to protect against wear. Furthermore, lubricants contribute to sealing against contaminants and are anticipated to guard against rusting and other contaminants' corrosive effects [1, 2]. They act in circulation systems as coolant, removing heat generated due to friction or heat conduction, e.g., to the bearing from the surrounding or external sources, and numerous lubricants are utilised as hydraulic fluids in devices within fluid transmission systems [6].

5.2.1 Types of Lubricants

There are an extensive variety of lubricants available for lubricating mechanical systems such as bearings and gears, machine elements and numerous biomechanical systems. However, the three principal lubricant types are liquid-oily, solid and gaseous lubricants [6]. *Animal and vegetable products*, which are one of examples of organic substance that constitute liquid-oily lubricants, were absolutely man's foremost lubricants and were utilised in huge amounts. Though due to their inadequate or lack of chemical inertness and certain lubrication essentials, *petroleum products* and *synthetic materials* have largely superseded them [6]. Several organic substances like lard and sperm oil, among others are nevertheless still being utilised as additives owing to their exceptional lubricating characteristics.

Lubricating with liquid-oily lubricants is comparable in several regards to lubricating with gaseous lubricants because similar fluid-film lubricating principles apply. Though both liquid-oily and gaseous lubricants are *viscous fluid lubricants*, they differ in two vital aspects—viscosity and compressibility [6]. The *viscosity of liquid-oily lubricants* is much greater than those of gaseous lubricants and their compressibility are much lower than for gaseous lubricants. Thus, *lubricating film thicknesses* and *load capacities* with liquid-oily lubricants are much greater compared with gaseous lubricants.

5.2.1.1 Liquid-Oily Lubricants

A *liquid-oily lubricant* is a fluid film of mostly liquid that comprises organic or inorganic substances used as lubricants in dry rubbing or sliding interface in relative motion. Examples of such organic or inorganic substances include mineral oils, synthetic esters, silicone fluids and water. The liquid-oily lubricants may be *conventional, synthetic,* or *a combination thereof* (i.e., *semisynthetic*). The *conventional lubricants*, commonly called petroleum lubricants are primarily hydrocarbon products exploited from naturally occurring fluids in the Earth. *Petroleum lubricants* are one of the most commonly utilised lubricants due to their blend of suitable properties such as *availability in desirable viscosities, low volatility, inertness* (i.e., resistance to lubricant deterioration), *corrosion protection* (i.e., resistance to corrosive deterioration of the lubricated contacts) and *affordability* [6]. Nevertheless, the inertness of petroleum lubricants is lost at elevated temperatures, like those experienced in modern *engines, causing the lubricant deterioration* through oxidation, and furthermore, leading to gums, varnishes and formation of other insoluble deposits. Consequently, if equipment longevity is desired, petroleum lubricants in almost all applications must be routinely replaced.

While the *synthetic lubricant* can typically be characterised as oily, neutral liquid materials, which is normally obtained indirectly from refined crude oil (i.e., petroleum) though possessing several comparable properties to lubricants obtained from petroleum. Synthetic lubricants are superior to lubricants derived from hydrocarbon products because they have relatively *superior viscosity stability with temperature changes* and a better *resistance to scuffing, oxidation* and *fire* [6]. Every type of synthetic lubricant has a specific purpose due to their vastly different properties. Consequently, not any single type of synthetic lubricant is fit for purpose for all lubrication applications. Therefore, all commercial synthetic lubricants (i.e., engine and gear oils) are a blend of numerous various synthetic lubricant types together with choice additives.

Grease is a form of liquid-oily lubricant that comprises oil or oils thickened with soap, soaps, or other thickeners into a solid or semisolid substance [6]. Greases can be made from unwholesome lard, waste animal's rendered fat, petroleum-derived or synthetic high viscosity oil. The grease of petroleum-derived or synthetic origin comprises thickening agents, which can be soap, an inorganic gel or an organic substance, dispersed in the liquid lubricant like the petroleum oil or a synthetic fluid. Thickening agents, such as soap, inorganic gel, or organic material, are scattered throughout the liquid lubricant, such as petroleum oil or synthetic fluid, whether it is petroleum-derived or synthetic in nature. Aluminium, barium, calcium, lithium, sodium and strontium soaps are the main thickening agents; non-soap thickeners include fine silica, modified clays and organic materials like phthalocyanine pigments. Other grease additives delay oxidation and corrosion, minimise friction, protect against wear and prevent viscosity change. When there are relatively large clearances between parts, the fluid component is the more significant lubricant; however, for smaller clearances, the molecular soap layers do the lubrication. Synthetic greases might comprise synthetic oils consisting of soaps or a blend of synthetic thickeners, or bases, in petroleum oils. They may be employed at a variety of temperatures and can be produced in both water-soluble and water-resistant forms. To obtain a unique feature, special-purpose greases may contain two or more soap bases or unique additives.

Greases are better lubricants than oils in situations where lubricant application is not required as frequently. They offer less lubricant dripping or splattering and, when desired, less sensitivity to inaccuracies in the mating parts because they function as a seal against lubricant loss and the ingress of contaminants.

5.2.1.2 Solid Lubricants

A *solid lubricant* is a solid film of solid substance that comprises organic or inorganic metallic compounds used as lubricants between rubbing or sliding interfaces in relative motion [6]. Examples of such organic or inorganic metallic compounds are graphite or molybdenum disulphide (MoS_2), respectively. Generally, there are *two broad classes* of solid organic lubricants. The two classes are *soaps, waxes and fats*, which include metallic soaps of calcium, sodium and lithium, *animal waxes* (like beeswax and spermaceti wax), *fatty acids* (like stearic and palmitic acids) and *fatty esters* (like lard and tallow); and *polymeric films*, which can be synthetic materials (like polytetrafluoroethylene (PTFE) and polychlorofluoroethylene). The predominant benefit of these film-type lubricants during exposure of their elements is their deterioration resistance. In contemporary pre-stressed concrete construction, for example, PTFE (13 mm) thick polymeric film plates are used to allow thermal movement of beams resting atop columns. The resistant polymeric film plate facilitates the structural members' expansion and contraction. If the adhesion to the substrate is good, thin films of soft metal on a hard substrate can function as effective lubricants. These metals include indium, lead and tin.

The three basic kinds of inorganic compounds that can be employed as solid lubricants are layer-lattice solids, miscellaneous soft solids and chemical conversion coatings [6]. The layer-lattice solid is any class of materials used as solid lubricants that have layers of crystal lattice structures with strong bonds between atoms in each layer and relatively weak bonds between atoms in different layers that allow the lamina to slide on top of one another. Examples of these materials are graphite and MoS_2. These materials, like graphite, are low-friction materials mainly because of their adsorbed films; without water vapour, graphite becomes abrasive and loses its lubricating qualities. Graphite and MoS_2 are both chemically inert and have high thermal stability. Layer-lattice solids such as tungsten disulphide, mica, boron nitride, borax, silver sulphate, cadmium iodide and lead iodide are other examples of solid lubricants that can be utilised.

The *miscellaneous soft solid* is a variety of inorganic solids that can be employed as solid lubricants. Examples of such inorganic solids include white lead, lime, talc, bentonite and silver iodide. The *chemical conversion coating* is a number of inorganic compounds employed as solid lubricants that are formed on metallic surfaces through chemical reaction. The commonly used chemical conversion coatings as lubricating film are substances such as chloride, oxalate, oxide, phosphate and sulphide films.

5.2.1.3 Gaseous Lubricants

A *gaseous lubricant* is a fluid film of gaseous materials that comprise gaseous substances used as lubricants in rubbing or sliding interface in relative motion [6]. Air, steam, industrial gases and liquid–metal vapours, among gases are just a few examples of the many gases that are utilised in this way, such as in gas bearings. Gaseous lubricant is often used to lubricate the rubbing or sliding surfaces of equipment handling various kinds of gases. This is done to prevent contamination of the lubricant and to make a particular device simpler to operate.

5.2.2 Properties of Lubricants

With the numerous organic or inorganic materials that are able to act as liquid, solid or exceptionally gaseous lubricants under specific conditions, a coverage of all their properties can be a daunting task or an impractical one. A mention of only those significant properties, generally considered as the properties of commercial fluid lubricants would be discussed. These properties include *viscosity, pour point, flash* and *fire point, oiliness, neutralisation number and penetration number*, among others.

5.2.2.1 Viscosity

Viscosity is the fluid's property that causes it to resist flowing under load. It's the *most significant* property of fluid lubricants because viscosity *influences the formation of EHD* or *thin films* between rolling elements and raceways or sliding interfaces of lubricated contacts, respectively, under both thin- and fluid-film conditions; governs the *extent of frictional effects encountered* between lubricated contacts of rolling or sliding surfaces; ascertains the *ease with which an engine at cold ambient conditions can be cranked* and whether an *adequate film might be formed* between bounding surfaces to *minimise friction and protect against wear* of solid-to-solid contact, among others [1, 2, 6, 7]. Customarily, viscosity is determined using a *viscometre* that measures the fluid lubricant's flow rate at standard conditions. The higher the fluid's flow rate in the viscometre, expressed in *seconds Saybolt universal* (SSU), *centipoise* (cP) or *reyns* based, respectively, on whether commercial unit, metric unit or English unit, the lower the viscosity [6]. In almost all liquids, viscosity diminishes considerably with an increase in temperature or at higher temperature. Because the little variation of viscosity with temperature fluctuations can be desirable to keep frictional changes at a minimal, lubricants are habitually rated with regards to viscosity index, and the minimal the viscosity change due to temperature fluctuations, the higher the viscosity index [8, 9].

5.2.2.2 Pour Points

The *pour point* of a fluid lubricant refers to the minimum temperature at which the fluid pours or flows when chilled without disturbance at prescribed conditions [6, 7]. Pour points can become the deciding factor when choosing a lubricant from a collection of lubricants with similar features since they play a significant role in evaluating flow properties at low temperatures [6]. Lubricants with low pour points are desirable because a low pour point suggests that the lubricant would function satisfactorily under cold operating conditions. This property of fluid lubricant is related to viscosity at low temperatures. Pour points are determined from the lowest temperature at which a particular flow is observed at a recommended, standard laboratory experiment. Almost all mineral oils are made up of multiple dissolved waxes. When these oils are chilled, the waxes start to crystallise (i.e., split into crystals) and fit together to form a rigid structure that traps the oil in tiny spaces within the structure. Once the waxed crystal structure is fully formed, the oil will stop flowing under controlled laboratory test conditions [7]. While the waxed crystal structure might be broken-up through mechanical agitation; however, it is possible to have oil flow at temperatures well below its pour point.

5.2.2.3 Flash and Fire Points

The *flash point* of a fluid lubricant refers to the temperature at which the fluid releases sufficient vapour at its surface to ignite when exposed to an open flame [6, 7]. For instance, as the temperature of lubricating oil in an open container rises, flammable vapours are released in increasing amounts. When the vapours' concentrations at the surface of the container become enormous enough, any exposure to an open flame causes the vapours to ignite and produce a brief flash [2]. Flash points, like the pour-point factor, may often play a significant role in choosing the right lubricant, in particular when lubricating machinery that handles extremely combustible materials [6]. A fluid's flash point serves as a general indicator of its flammability or combustibility; below the flash point, there is an inadequate vapour present to enable combustion, but at a certain temperature above it, the fluid produces adequate vapour to do so. This temperature above the flash point that the fluid produces adequate vapour to support combustion is termed '*fire point*' [6].

5.2.2.4 Oiliness

The *oiliness* of a fluid lubricant refers to the fluid's tendency to wet and adhere to surfaces [6]. Generally, oiliness denotes the comparative capability of a lubricating oil to function under boundary lubrication conditions. There is not yet any established standard test for measuring the oiliness of a lubricant, as the primary methods for determining oiliness are experience and subjective judgement [6]. It is not always necessary for the best fluid lubricant for a given application to be the oiliest; for instance, packed rolling bearings usually benefit from long-fibre grease that is less oily than lubricating oils.

5.2.2.5 Neutralisation Number

The *neutralisation number* of a fluid lubricant refers to a measure of the acid or alkaline content of the new fluid or an indicator of the spent fluid's degree of oxidation degradation [6]. The value of the neutralisation number of a fluid lubricant is determined by titration, an analytic standard chemical method defined as the number of alkalines (i.e., potassium hydroxide) in milligrams needed to neutralise one gram of the lubricant.

5.2.2.6 Penetration Number

As applied to grease, *penetration number* refers to a measure of the grease's film characteristics [6]. An established standard test that involves dropping a standard cone into a sample of the grease under test is used to determine the value of a grease's penetration number. The gradations on the cone show how deep the penetration is, and the deeper the number, the more fluid the grease contains.

5.2.3 Deterioration of Lubricants

The continuous use of the same fluid lubricant leads to *deterioration of the lubricant quality*. For example, the continuous use of the same lubricating oil leads to the deterioration of the lubricating oil quality over a period of time, even when centrifuge purification techniques are employed to clean it, such as in internal combustion engines. The backbone of lubricating oils, i.e., the base oils, which may be *conventional, synthetic*, or *a combination thereof* (i.e., *semisynthetic*), is the component in lubricated contacts that is responsible for minimising friction and protecting against wear [7, 8, 10]. For the duration of the oil's life, almost all of the oil additives to the base oil are designed to be sacrificial and consumed. Thus, like everything else, lubricants have a lifespan, and they might deteriorate due to *depletion of the additives over time, loss of viscosity, becoming contaminated, or a combination thereof* [11].

The reasons the quality or effectiveness of the lubricating oil is lost over time, or the causes of oil degradation that make it require replacement after some time of usage, include *viscosity loss, oxidation, thermal breakdown, additive depletion, condensate formation* and *fuel dilution* [11, 12].

5.2.3.1   Viscosity Loss

The viscosity of lubricating oil is its most important property, having a direct bearing on minimising friction and protecting against wear [1, 2, 13]. The intense pressure the lubricating oils bear as they are squeezed between opposing lubricated contacts in motion might shear or disintegrate their molecular structures, causing *viscosity loss* [14]. Engines are designed to operate best using a lubricating oil of a specific viscosity and any viscosity loss might result in increased friction and the required engines' wear protection may be compromised.

5.2.3.2   Oxidation

The interaction between molecules of lubricating oil and oxygen molecules, in general, causes chemical breakdown [13]. Oxygen in the presence of moisture causes exposed metal to rust and oxidise, and it breaks down the base oil of the lubricating oil into harmful deposits and sludge, reducing the oil's effectiveness [2, 13]. A multitude of factors influence the lubricating oil's oxidation rate, which rises rapidly with temperature and also oxidises more quickly when exposed to oxygen contained in the surrounding air or when the oil is mixed more closely with the oxygen. The oxidation may cause a reduced or increased lubricating oil viscosity that negatively impacts its operational and energy efficiencies.

5.2.3.3   Thermal Breakdown

Engines, in particular modern ones, operate at temperatures as high as 113°C, and sometimes even higher, which can cause the lubricating oil's base oil to break down. Generally speaking, the rate at which base oil oxidises doubles for every 10°C increase in temperature, causing the lubricating oil to corrode and oxidise [2]. The oil's lifespan can be extended, and the *thermal breakdown* response can be reduced by keeping it as cool as possible while in use [13].

5.2.3.4   Additive Depletion

As a sacrificial measure, the majority of oil additives are designed to deplete during normal use or while ageing during the oil's lifespan [12, 14]. Lubricating oil *additive depletion* typically occurs in two forms; both are frequent and can happen at the same time [12]. The additive depletion can be caused by *decomposition*, in which the additive mass strays in the oil but its molecular structure changes, forming a range of transformed products; i.e., different molecules. However, the transformed products, in certain occurrences, might have properties the same as the initial additives, but in almost all circumstances, these properties are typically lost or degraded; and *mass transfer*, which takes place when the additives transfer mass during normal operation, typically due to the additive fulfilling its intended sacrificial role [12].

5.2.3.5 Condensate Formation

Engines are compromised by temperature fluctuations, even when not in use, and those fluctuations lead to the *formation of condensate*, causing water contamination in the engine [14]. Leaving an engine unused for extended periods or used for short periods that do not permit full warm-up of the engine allows water to be retained in the oil rather than evaporating and exiting through the tailpipe. The presence of moisture might cause the formation of sludge, and it not only degrades the lubricating oil but also gives rise to bacteria and fungus growth, causing the lubricating oil to corrode and oxidise. These drastically change the quality and properties of lubricating oil, making it useless.

5.2.3.6 Fuel Dilution

Lubricating oil can become contaminated and lose its viscosity when fuel washes past the piston rings. Frequent short trips that do not permit the lubricating oil to attain average operating temperature may, in particular, be problematic because the fuel will never volatilise and exit. Excessive *fuel dilution* of lubricating oil causes the formation of sludge and varnish, necessitating lubricating oil replacement more often [14].

5.2.4 Lubricant Selection

In general, all mechanical systems, including internal combustion engines, rolling element bearings and gears, as well as machine elements require premium-quality lubricants, many of which are specially formulated for this use and various applications.

5.2.4.1 Oil Selection

The characteristics that any selected lubricating oil should possess to effectively lubricate rolling element bearings, gears and machine elements include *excellent oxidation resistance at working temperatures* to offer a long lifespan without thickening or deposit formation that obstructs bearings, gears or machine elements operation; *suitable viscosity at operating speeds and temperatures* to minimise friction and protect against wear of bearings, gears or machine elements; *antirust properties* to prevent rusting while moisture is present in bearings, gears or machine elements; *good antiwear properties* where necessary due to heavy or extreme loads in bearings, gears or machine elements; and *excellent demulsibility to enable water separation* in circulation systems [1, 2, 15]. The association between lubricating oil properties and fatigue is currently receiving more attention in addition to these qualities. It is difficult to pinpoint the precise lubricant characteristics that affect fatigue, but where average bearing, gear or machine elements' life is less than anticipated, it may be advantageous to use specialised lubricants designed to reduce fatigue. The antifatigue qualities of lubricants can be improved by combining particular additives with particular base stocks.

5.2.4.2 Grease Selection

To successfully lubricate rolling element bearings, gears and machine elements, any selected lubricating grease should possess characteristics that include *good resistance to oxidation* to prevent deposit formation or hardness that can reduce bearings, gears, or machine elements' lives. Grease employed in '*packed for life*' bearings, gears or machine elements is anticipated to offer trouble-free operation for the lifespan of a mechanical system, given typical operating conditions; *mechanical stability to inhibit softening or hardening excessively* due to shearing while in use; *appropriate feedability at operating temperatures and consistency* for the application method required to maintain enough lubricating film thickness without excessively slumping, that causes rising friction; *controlled grease bleeding*, in particular, for high-speed bearings, to provide the tiny amount of lubrication required for EHD lubricating films formation; *improved antiwear qualities* to withstand the rubbing action between roller ends and raceway flanges in bearings carrying significant radial or axial thrust loads; and *excellent capacity to prevent rusting* on surfaces [1, 2, 15]. In the event that a bearing, gear, or machine element becomes contaminated with small amounts of water, the grease must be able to absorb the water without becoming noticeably softer or harder, be compatible with systems and their components, and, depending on the application, be resistant to corrosion or deterioration in situations where small amounts of acids or caustic materials could get into the grease.

5.2.5 *Automotive Gear Lubricants*

All automobiles, ancient and modern alike, have needed some sort of gearing for the power from any engine to be transmitted to the driving wheels. With this gearing, there are various gear designs, including spur, helical, herringbone and hypoid gears, with all gears requiring lubrication. A wide range of performance levels are available to suit mild to severe operating and application conditions, much as there is a wide range of gearing and application requirements.

Based on the required performance metrics, finished gear oils are usually made up of premium-quality base stocks (minerals or synthetics) and 5% to 20% additives [2]. Gear oils can be formulated with up to 10 various additives; however, more additives might be required due to the growing demands of extended service intervals and environmental concerns. Examples of these additives are metal deactivators, oxidation inhibitors, pour-point depressants, defoamants, rust and corrosion inhibitors, antiwear agents, detergents and dispersants, and viscosity index (VI) improvers, among others [2]. Like other high-performance lubricants, these additives must be balanced to meet the necessary performance requirements because they compete with one another to fulfil their respective roles.

To specify the lubricant requirements for automobile gear, three predominant learned societies, i.e., the Society of Automotive Engineers (SAE), the American Society for Testing and Materials (ASTM) and the American Petroleum Institute (API), have teamed up with producers of additives, formulators of lubricants, and equipment manufacturers and users. Automotive gear lubricants are classified according to their viscosity using a viscosity classification system (SAE J306) established by SAE, while ASTM defines test limits and performance level ratings employed for test procedures and standards, and API defines the language employed for performance categories [2, 9]. Apart from SAE, ASTM and API, the US military established MIL-PRF-2105E, a widely employed specification for gear lubricants. There are currently no licencing regulations for gear oils by API, in contrast to automobile engine oils [2]. Nonetheless, a few significant OEMs grant permission to employ their names for axle and gearbox lubricants.

The following are the performance categories for the API [2]:

API GL-1: Lubricants intended for use in manual transmissions under moderate service circumstances. Pour-point depressant, oxidation and corrosion inhibitor and antifoam agent additives are present in these oils, but extreme pressure (EP), antiwear agent and friction modifier additives are absent.

API GL-4: Lubricants intended for use in differentials with hypoid gearing or spiral bevel gearing under moderate to severe operating circumstances. In certain manual transmissions and transaxles where EP oils are appropriate, these oils may be used.

API GL-5: Lubricants designed for differentials with hypoid gears that are subjected to extreme torque and sporadic shock loading. High concentrations of antiwear and EP additives are typically present in these oils.

API MT-1: Lubricants designed for manual transmissions without synchronisers. GL-1, GL-4 and GL-5 API products do not have the same levels of oxidation and thermal stability as these oils because of their formulation.

5.2.5.1 Automatic Transmission Fluids

Among the most technologically advanced lubricants in the market are automatic transmission fluids. Automatic transmission fluid serves as the medium for heat and power transfers in the converter section, lubricates gears and bearings in the gearbox, regulates the frictional properties of clutches and bands, and functions as hydraulic fluid in control circuits [2]. For prolonged service periods, all of these functions must operate satisfactorily across a wide temperature range, from the lowest anticipated ambient temperatures to operating temperatures of 300°F (149°C) or higher. Of course, a fluid cannot be deemed appropriate for this kind of service until it has undergone a thorough evaluation.

The big American automakers, including Ford, DaimlerChrysler and General Motors, are still working to develop better automatic transmission fluids. Increases in oxidation stability, antiwear retention, shear stability, low-temperature fluidity, material compatibility and fluid friction stability are required because the improvement is intended for fill-for-life applications (100,000 to 150,000 miles) [2].

5.3 Additives

For the purpose of achieving a balanced blend of performance characteristics in the finished lubricants, the *basic stock of lube oil* serves *as the building block* or basis upon which suitable additives are chosen and blended [1]. It should be emphasised that distinct base stock manufacturing techniques can also produce base stocks with the appropriate characteristics to produce premium finished lubricants with the right performance levels. To maximise the performance of finished lubricants, it is critical to comprehend how base stocks and additives interact and to match those requirements to the needs of machinery and potential operating environments.

An *additive* is a compound, either organic or inorganic, that is suspended as solids or dissolved in lubricating oils to impart specific properties to the finished oils [2, 16–19]. Generally, based on the engine, additives can range from 0.1% to 30% of the oil volume. While some additives impart the lubricant with new and beneficial properties, others enhance existing properties and work to slow down the rate at which the product undergoes undesirable changes over the course of its service life [17, 19]. Lubricating oil additives have significantly contributed to the advancement of industrial machinery and prime movers by enhancing their performance attributes [17]. The development of many modern equipment types, such as hypoid gears, high-speed gas and steam turbines, railroad and marine diesel engines, and several others, would not have happened as quickly in the absence of additives and the performance advantages they offer.

5.3.1 Roles of Additives and Functions Within Lubricants

Understanding the roles and functions of additives within lubricants is crucial because there is more to lubricants than just viscosity. There are three basic roles of additives. The roles are to *enhance the properties of base oils* by adding pour-point depressants, antioxidants and corrosion inhibitors; *suppress the undesirable properties of base oils* by adding viscosity index (VI) improvers and pour-point depressants; and *impart the base oils with new properties* by adding EP additives and detergents [17, 19, 20].

5.3.2 Commonly Used Additives

Among the more commonly used additives are pour-point depressants, defoamants, VI improvers, antiwear additives and EP additives [18, 19]. While some of these additives serve multiple functions, like VI improver, which equally serves as a pour-point depressant, dispersant, or antiwear agent, and it is, however, employed only for its primary role.

5.3.2.1 Pour-Point Depressants

A *pour-point depressant* is a high-molecular-weight polymer that functions by preventing wax crystal structure formation, which otherwise prevents the flow of oil at low temperatures [18, 19, 21]. There are two pour-point depressant types that are commonly employed: polymethacrylates, which cocrystallise with the wax to inhibit crystal growth and alkylaromatic polymers, which are adsorbed on the wax crystals during their formation to prevent them from growing and sticking to one another [17]. While pour-point depressants decrease the temperature at which rigid structures are formed, they do not completely prevent the growth of wax crystals.

5.3.2.2 Viscosity Index Improvers

A *viscosity index improver* is a long-chain, high-molecular-weight polymer that functions by elevating the oil's relative viscosity at higher temperatures than it does at lower temperatures [17, 20, 22]. A temperature increase in the mixture typically results in a modification of the polymer's physical configuration. It is speculated that the polymer's molecules take on a coiled shape in cold oil to reduce their impact on viscosity. The molecules in hot oil have a tendency to straighten up, and this contact with the oil results in a thickening effect that is proportionately stronger. It should be noted that while the viscosity of the oil-polymer mixture drops with increasing temperature, it does so less dramatically than it would have in the oil alone. Methacrylate polymers and copolymers, acrylate polymers, and olefin polymers and copolymers are some of the main VI improvers. The polymer's molecular weight distribution determines how much VI is improved in these materials. Hydraulic fluids, automatic gearbox fluids and engine oils all employed VI improvers. They are also components in car gear lubricants. Their application enables the creation of formulations that, when combined with straight mineral oils alone, offer adequate lubrication across a far larger temperature range.

5.3.2.3 Defoamants

A *defoamant* is an additive that helps lubricating oils resist foaming by degassing or deaerating foam [2]. The crude oil types, viscosity, and type and degree of refining all have a significant impact on the degree of foaming resistance. While the oil may have a strong tendency to foam and become agitated in some applications, in other situations, even a small amount of foam can cause serious problems; in these situations, a defoamant may be added to the oil. Defoamants are classified into defoamers, which are post-additives that reduce foam that has already formed, and antifoams, which stop foam from forming in the first place and can manage foam for a set amount of time [17].

5.3.2.4 Oxidation Inhibitors

An *oxidation inhibitor* is a substance used to slow down or stop the oxidation that happens when oil is heated in the presence of air [2]. Oxidation, which occurs when oil is heated in the presence of air, increases the oil's viscosity and organic acid concentration. Furthermore, hot metal surfaces exposed to oil may form deposits of lacquer and varnish, which, in extreme cases, may further oxidise to produce hard, carbonaceous materials. To slow down or stop the oxidation, a variety of materials, including metals, in particular copper and iron, and organic and mineral acids, can function as catalysts or oxidation inhibitors [18].

5.3.2.5 Rust and Corrosion Inhibitors

A *rust and corrosion inhibitor* is a film that can be adsorbed or chemically bonded to a metal surface to form a protective barrier that keeps corrosive substances from penetrating or damaging the metal [17, 23]. The majority of compounds employed as rust inhibitors possess a powerful polar attraction to the surfaces of metals [20]. Often, the cause of this attraction is physical or chemical interactions at the metal's surface, where the compounds mix to form a thick layer that keeps water from passing through [18, 19]. Alkaline earth sulfonates and amine succinates are common materials used for this purpose. The two most common causes of corrosion in lubricating oil systems are most likely the organic acids that build up in the oil and the contaminants that the oil transfers and absorbs. Because the organic acids in the oil easily corrode some metals, like lead in the copper-lead or lead-bronze employed in these inserts, the bearing inserts used in internal combustion engines, for example, may suffer corrosion [18, 19]. The corrosion inhibitors form a protective film on the bearing surfaces, which protects the metal from corrosive agents [21]. As a result, rust and corrosion are inhibited.

5.3.2.6 Detergents and Dispersants

A *detergent and dispersant* is an additive that either inhibits the oxidative breakdown of lubricating oil, retards the formation of deposits, or minimises the accumulation of deposits on metal surfaces [2, 21]. To control deposit formation, detergents and dispersants are employed to inhibit oxidative breakdown or suspend the already-formed hazardous compounds in the oil. Most of the time, detergents and dispersants take care of the suspending part, and oxidation inhibitors inhibit the oxidation mechanism [23]. Detergents, which are organic acids of metallic salts that often include excess bases associated with them, typically in the carbonate form; and dispersants, which are metal-free compounds with larger molecular weights than detergents; complement each other as additives [20].

Many factors can contribute to oil deterioration and hazardous deposit formation when using internal combustion engines. These deposits have the potential to disrupt oil flow, build up behind piston rings, result in ring sticking and quick wear, affect clearances, and hinder the proper functioning of essential parts. For deposits to either not form at all or accumulate more slowly on metal surfaces, the oil needs to contain detergent and dispersant [17]. These deposits are typically difficult to remove from metal surfaces once they have formed and accumulated, unless mechanical cleaning is used. So, regularly draining and replacing the oil is crucial because it keeps the impurities in the oil out of the engine prior to the oil's capacity for containment being exceeded.

5.3.2.7 Antiwear Additives

An *antiwear additive* is an additive that is used to minimise friction, protect against wear, and prevent scuffing and scoring during boundary lubrication or when lubricating fluid films cannot be maintained [2]. With increasing loads or temperatures, the opposing asperities make contact and cause the oil film to become thinner, increasing friction, which can lead to welding. The welds break instantly as the sliding process goes on, but wear particles and metal transfer can also create new roughness that leads to scuffs and scores. On the basis of the degree of requirement, two basic material classes—*mild antiwear and friction-reducing additives*, also referred to as boundary lubrication additives are employed to prevent contact between opposing surface asperities. Their examples are polar materials like acids, esters and fatty oils.

5.3.2.8 Extreme Pressure Additives

An *extreme pressure additive* is an additive that is required at high temperatures or under heavy loads where sliding conditions are more severe to minimise friction, protect against wear and prevent severe surface damage [14]. By thermochemically reacting with metal surfaces, EP additives form incredibly strong protective films that eliminate direct contact between surface asperities and prevent scoring and seizing. These films are capable of withstanding temperature and pressure extremes. The reaction's kinetics are determined by the surface temperatures produced by the localised high temperatures caused by the contacts of opposing asperities and the breaking of junctions between these asperities [2].

5.4 Tribological Aspects of Slides, Guides and Ways

The category of bearings, also referred to as flat bearings, includes all slides, guides and ways employed in stamping and forging presses, and the crosshead guides of certain compressors, diesel engines and steam turbines, as well as on machine tools like lathes, grinders and milling machines [2, 14]. These bearings work in a wide range of service situations. Crosshead guides run at comparatively fast speeds in circumstances that frequently allow fluid-lubricating films to form [14]. The guides are typically intended to work on the same lubricating fluid employed for the machine's primary and connecting rod bearings; therefore, the needs for lubrication are typically not severe, but special challenges might arise with the lubrication of machine tool ways and slides [1, 2, 14]. The lubricant has a tendency to be wiped off at slow speeds and under large loads, which makes boundary lubrication prevail. Although there is more friction as a result, boundary films benefit from having a nearly constant thickness. For the slide to be lifted and floated at moderate loads and quick traversal speeds, the oil's viscosity must be high enough to allow fluid films to form. These fluid films' thickness can fluctuate significantly with changes in speed or load, which can result in the items being machined having wavy surfaces or causing them to run out of size. Hence, the slides and ways must always work under boundary conditions as a result of precision machining. This typically requires the oil to contain additives that reduce friction and protect against wear.

When two items glide over one another, a phenomenon known as the *stick–slip phenomenon*, which causes a sudden jerking action, can be observed. More force is needed to start the slide from rest than to keep it moving once it has started if the lubricant's static coefficient of friction is higher than its dynamic coefficient. When a force is applied to start the slide, there will initially be resistance because the feed mechanism has some inherent free play [1, 2]. The slide will start to move if there is a sufficient amount of force, and once the slide starts moving, the effort needed to keep it in motion reduces as it advances until the feed mechanism uses up the free play. This can happen continuously at slow traverse speeds, leaving chatter marks on the workpiece. The stick–slip phenomenon can make it very challenging to precisely establish feed depths for cross-slides. By using additives to bring the static coefficient of friction down to a value equal to or lower than the dynamic coefficient, stick–slip phenomena can be prevented [1].

The lubricant has a tendency to drain from the surfaces of vertical guides. Special adhesive qualities are required to thwart this tendency and secure appropriate films. Based on the aforementioned requirements, appropriate lubricants for slides, guides and ways should possess characteristics that include *a suitable viscosity at the operating temperature for easy distribution* to the sliding surfaces and the formation of the required boundary films; *high film strength to maintain the required boundary films* under heavy loads and antiwear ability to control wear under these boundary conditions; and *appropriate frictional properties to protect against wear* under these boundary conditions [1, 14]. Some devices, such as forging machines and open-crankcase steam engines, may expose their slides and guides to a lot of water. Lubricants for these applications must therefore be specially made to prevent washing off.

5.5 Rolling Element Tribology and Applications

The concept of rolling motion is applied to all wheel types, rolling element bearings and different transmission drives, including gears, belt drives and cam followers [24, 25]. A wheel and its track, rolling element bearings, or gears frequently exhibit tangential traction in the contact zone, or they exhibit some degree of slip [24, 25]. The automotive industry, electric motors, machine tools, conveyors, escalators, gyroscopes, aircraft, ships, etc. are just a few examples of where rolling element bearings and gears have found applications. Rolling element bearings and gears have a long operational life because, under typical lubricating conditions, friction and wear are very minimal. The tangential and contact stresses are significantly increased when there is no lubricating film present between opposing contacts, which shortens the mechanical system's lifespan [24].

5.5.1 Rolling Element Bearings

The physics of the rolling element bearings led to *rolling bearing tribology*. In other words, the geometry, kinematics, contact stresses and materials used in rolling bearings all affect the friction and wear they encounter. A bearing's function is to transmit load between a shaft and a housing while allowing for relative positioning and rotational freedom [25]. Rolling elements can be inserted between the moving surfaces to transmit loads between surfaces that are moving relative to one another in mechanical systems. The significantly less friction associated with rolling then takes the place of the greater friction associated with sliding. To suit the varied combinations of loads and speeds necessary in mechanical applications, numerous rolling element bearing types and their lubricating fluid's viscous resistance have been established [25].

The state of the lubricating fluid's viscous resistance during normal operating conditions has a major influence on the rolling bearing's tribological performance [25, 26]. The notion of viscous flow can be employed in applying the lubrication theory to viscous lubricating fluids in a rolling bearing. Between the stationary and rotating shaft surfaces, a very thin layer of lubricating fluid is maintained. Conversely, a light oil might not be capable of maintaining the necessary coating between bounding surfaces, which could lead to surface wear. A very viscous oil increases resistance and results in significant power loss. Thus, expressions for the power absorbed due to the lubricating fluid's viscous resistance in journal, foot-step and collar bearing types can be generated for various bearing applications [26].

5.5.1.1 Journal Bearings

Consider a *journal bearing* in which its annular space between the rotating shaft and the bearing is filled with lubricating fluid of viscosity μ. If the diameter of the shaft is D, lubricating fluid-film thickness is t, bearing length is L, and the speed of the shaft is N; the power absorbed in overcoming the viscous resistance due to the viscosity of the lubricating fluid in the bearing can be derived as follows [26]:

Assuming the angular speed of the shaft (in rad/s), $\omega = \frac{2\pi N}{60}$; the tangential speed of the shaft (in m/s), V can be expressed using Eq. 5.1:

$$V = \omega R \rightarrow V = \frac{2\pi N}{60} \times \frac{D}{2} = \frac{\pi D N}{60}, \tag{5.1}$$

where R denotes the radius of the shaft. If the thickness of the film t of the lubricating fluid is light, a linear velocity distribution, $\frac{du}{dy}$ can be assumed, i.e., $\frac{du}{dy} = \frac{V}{t} = \frac{\pi D N}{60t}$.

Therefore, the shear stress, τ, can be expressed using Eq. 5.2:

$$\tau = \mu \frac{du}{dy} = \frac{\mu \pi D N}{60t}. \tag{5.2}$$

The shear force or viscous resistance, F is the product of the shear stress, τ (on the lubricating oil by the shaft) and the shaft's surface area $\pi D L$, i.e.,

$$F = \frac{\mu \pi D N}{60t} \times \pi D L = \frac{\mu \pi^2 D^2 N L}{60t}. \tag{5.3}$$

Therefore, the torque, T needed to overcome the viscous resistance of the lubricating fluid in a journal bearing can be expressed using Eq. 5.4:

$$T = F \times \frac{D}{2} = \frac{\mu \pi^2 D^2 N L}{60t} \times \frac{D}{2} = \frac{\mu \pi^2 D^3 N L}{120t}. \tag{5.4}$$

Hence, power (in Watts) absorbed in overcoming the viscous resistance of oil in the bearing, P can be expressed using Eq. 5.5:

$$P = \omega T \rightarrow P = \frac{2\pi N}{60} \times \frac{\mu \pi^2 D^3 N L}{120t} = \frac{\mu \pi^3 D^3 N^2 L}{3,600t}. \tag{5.5}$$

Example 5.1 A 200 mm-long journal bearing houses a 100 mm-diameter shaft that revolves at 60 rpm. If the two surfaces are evenly spaced apart by 0.5 mm and the lubricating oil has a linear velocity distribution with a dynamic viscosity of 0.04 poise, determine the.

(i)   Amount of torque necessary to overcome the oil's viscous resistance in the journal bearing and.
(ii) Power absorbed in the journal bearing.

Solution

Given: Diameter of the shaft, $D = 100$ mm $= 0.1$ m; Speed of the shaft, $N = 60$ rpm; Length of the bearing, $L = 200$ mm $= 0.2$ m; Thickness of oil film, $t = 0.5$ mm $= 0.0005$ m; and Dynamic viscosity, $\mu = 0.04$ poise $= 0.004$ Ns/m^2.

(i)   The amount of torque required to overcome the viscous resistance of the oil in the journal bearing, $T = ?$

Therefore, the torque required to overcome the oil viscous resistance in the journal bearing, $T = \frac{\mu \pi^2 D^3 N L}{120t}$

$$= \frac{0.004 \times \pi^2 \times 0.1^3 \times 60 \times 0.2}{120 \times 0.0005}$$
$$= 0.00789 \text{ Nm.}$$

(ii) Power absorbed in the journal bearing, $P = ?$

Therefore, the power absorbed in the bearing, $P = \omega T = \frac{2\pi N}{60} T$

$$= \frac{2\pi \times 60 \times 0.00789}{60}$$
$$= 0.0496 \text{ W}$$
$$= 0.05 \text{ W.}$$

Consider a *foot-step bearing* at the end of a vertical shaft in which the space between the surface of the shaft and the bearing is filled with lubricating fluid of viscosity μ. If the diameter of the shaft is D, length of the bearing is L, oil film thickness is t, and the speed of the shaft is N; the power absorbed in overcoming the viscous resistance due to the viscosity of the lubricating fluid in the bearing can be derived as follows [26]:

Assuming an elementary circular ring of radius, r and thickness, dr; hence, the area of the elementary ring, A_{rg} can be expressed using Eq. 5.6:

$$A_{rg} = 2\pi r dr. \tag{5.6}$$

The viscous shear stress, τ of the elementary ring can be expressed using Eq. 5.7:

$$\tau = \mu \frac{\delta u}{\delta y} = \mu \frac{V}{t}, \tag{5.7}$$

where V denotes the tangential velocity of the shaft at radius, r and it can be expressed using Eq. 5.8:

$$V = \omega r = \frac{2\pi N r}{60}. \tag{5.8}$$

Therefore, the shear force of the elementary ring, dF, defined as the product of the viscous shear stress, τ of the elementary ring and its area A_{rg}, i.e.,

$$dF = \tau A_{rg} = \mu \frac{V}{t} 2\pi r dr = \mu \times \frac{2\pi N r}{60t} \times 2\pi r dr = \mu \times \frac{\pi^2 N r^2}{15t} dr. \tag{5.9}$$

The torque, dT on the ring is the product of the shear force, dF of the elementary ring and its radius, r, i.e.,

$$dT = dF \times r = \frac{\mu \pi^2 N r^2}{15t} dr \times r = \frac{\mu \pi^2 N r^3}{15t} dr. \tag{5.10}$$

Thus, the torque, T required to overcome the viscous resistance of the oil in the foot-step bearing can be expressed using Eq. 5.11:

$$T = \frac{\mu \pi^2 N}{15t} \int_0^R r^3 dr = \frac{\mu \pi^2 N}{15t} \left[\frac{r^4}{4} \right]_0^R = \frac{\mu \pi^2 N R^4}{60t}. \tag{5.11}$$

Hence, power (in Watts) absorbed in overcoming the viscous resistance of oil in the bearing, P can be expressed using Eq. 5.12:

$$P = \omega T \rightarrow P = \frac{2\pi N}{60} \times \frac{\mu \pi^2 N R^4}{60t} = \frac{\mu \pi^3 N^2 R^4}{1,800t}. \tag{5.12}$$

Example 5.2 A foot-step bearing at its lower end supports the 100 mm-diameter vertical shaft's lower end, which revolves at 750 rpm. If the flat end of the shaft and the bearing's surface are separated by a 0.5 mm-thick oil coating at a dynamic viscosity of 1.5 poise, calculate the.

(i) Amount of torque needed to overcome the oil's viscous resistance in the foot-step bearing and.
(ii) Power needed to rotate the vertical shaft.

Solution

Given: Diameter of the shaft, $D = 100 \, \text{mm} = 0.1 \, \text{m}$; Speed of the shaft, $N = 750 \, \text{rpm}$; Thickness of oil film, $t = 0.5 \, \text{mm} = 0.0005 \, \text{m}$; and Dynamic viscosity, $\mu = 1.5 \, \text{poise}$ $= 0.15 \, \text{Ns/m}^2$.

(i) Amount of torque needed to overcome the viscous resistance of the oil in the foot-step bearing, $T = ?$

Therefore, the torque needed to overcome the oil viscous resistance in the foot-step bearing, $T = \frac{\mu \pi^2 N R^4}{60t}$

$$= \frac{0.15 \pi^2 \times 750 \times \left(\frac{0.1}{2}\right)^4}{60 \times 0.0005}$$
$$= 0.2313 \, \text{Nm}.$$

(ii) Power needed to rotate the vertical shaft, i.e., absorb in the bearing, $P = ?$

Therefore, the power absorbed in the bearing, $P = \frac{2\pi N}{60} T$

$$= \frac{2\pi \times 750 \times 0.2313}{60}$$
$$= 18.16 \, \text{W}$$

5.5.1.2 Collar Bearings

Consider a *collar bearing* that withstands axial load from a shaft, where the face of the collar is separated from the surface of the bearing with lubricating oil of viscosity μ, generating a lubricating film of uniform thickness that is maintained by a forced lubrication system. If the internal radius of the collar is R_1, external radius of the collar is R_2, oil film thickness is t and the speed of the shaft is N; the power absorbed in overcoming the viscous resistance due to the viscosity of the lubricating oil in the bearing can be derived as follows [26]:

If an elementary circular ring of radius r and thickness dr; the area of the elementary ring, A_{rg} can be expressed using Eq. 5.13:

$$A_{rg} = 2\pi r \, dr. \tag{5.13}$$

The viscous shear stress, τ of the elementary ring can be expressed using Eq. 5.17:

$$\tau = \mu \frac{\delta u}{\delta y} = \mu \frac{V}{t}, \tag{5.14}$$

where V denotes the tangential velocity of the shaft at radius r, and it can be expressed using Eq. 5.15:

$$V = \omega r = \frac{2\pi N r}{60}. \tag{5.15}$$

Therefore, the shear force of the elementary ring, dF, defined as the product of the viscous shear stress, τ of the elementary ring and its area A_{rg}, i.e.,

$$dF = \tau A_{rg}. \tag{5.16}$$

Substituting the values of viscous shear stress, τ and area of the elementary ring, A_{rg} into Eq. 5.16, i.e.,

$$dF = \tau A_{rg} = \mu \frac{V}{t} 2\pi r \, dr. \tag{5.17}$$

Similarly, substituting the value of tangential velocity, V into Eq. 5.17, i.e.,

$$dF = \mu \frac{V}{t} 2\pi r \, dr = \mu \times \frac{2\pi N r}{60t} \times 2\pi r \, dr = \mu \times \frac{\pi^2 N r^2}{15t} dr. \tag{5.18}$$

The torque, dT on the ring is the product of the shear force, dF of the elementary ring and its radius, r, i.e.,

$$dT = dF \times r. \tag{5.19}$$

Substituting the value of the shear force, dF of the elementary ring into Eq. 5.19, i.e.,

$$dT = dF \times r = \frac{\mu \pi^2 N r^2}{15t} dr \times r = \frac{\mu \pi^2 N r^3}{15t} dr. \tag{5.20}$$

Thus, the torque, T required to overcome the viscous resistance of the oil in the foot-step bearing can be expressed using Eq. 5.21:

$$T = \frac{\mu \pi^2 N}{15t} \int_{R_1}^{R_2} r^3 dr = \frac{\mu \pi^2 N}{15t} \left[\frac{r^4}{4} \right]_{R_1}^{R_2} = \frac{\mu \pi^2 N \left(R_2^4 - R_1^4 \right)}{60t}. \tag{5.21}$$

Hence, power (in Watts) absorbed in overcoming the viscous resistance of oil in the bearing, P can be expressed using Eq. 5.22:

$$P = \omega T. \tag{5.22}$$

Substituting the values of angular velocity of the shaft, ω and the torque, T into Eq. 5.23, i.e.,

$$P = \frac{2\pi N}{60} \times \frac{\mu \pi^2 N (R_2^4 - R_1^4)}{60t} = \frac{\mu \pi^3 N^2 (R_2^4 - R_1^4)}{1800t}. \tag{5.23}$$

Example 5.3 To support a shaft's thrust at 300 rpm, a collar bearing with internal and external dimensions of 180 mm and 240 mm is employed. If a lubricating coating with a thickness of 0.25 mm and a viscosity of 0.8 poise is maintained between the collar and the bearing, calculate the.

(i) Amount of torque needed to overcome the oil's viscous resistance in the collar bearing and.
(ii) Power loss in overcoming the oil's viscous resistance.

Solution

Given: Internal radius of the collar, $R_1 = 180/2 = 90$ mm $= 0.09$ m; External radius of the collar, $R_2 = 240/2 = 120$ mm $= 0.12$ m; Speed of the shaft, $N = 300$ rpm; Thickness of the oil film, $t = 0.25$ mm $= 0.00025$ m; and Viscosity of the oil, $\mu = 0.8$ poise $= 0.08$ Ns/m^2.

(i) Amount of torque needed to overcome the oil's viscous resistance in the collar bearing, $T = ?$

Therefore, the torque required to overcome the viscous resistance of the oil in the collar bearing, $T = \frac{\mu \pi^2 N (R_2^4 - R_1^4)}{60t}$

$$= \frac{0.08\pi^2 \times 300(0.12^4 - 0.09^4)}{60 \times 0.00025}$$

$$= 2.238 \text{ Nm}.$$

(ii) Power loss or absorbed in overcoming the oil's viscous resistance, $P = ?$

Therefore, the power loss or absorbed in overcoming the viscous resistance of the oil, $P = \frac{2\pi N}{60} T$

$$= \frac{2\pi \times 300 \times 2.238}{60}$$

$$= 70.31 \text{ W}.$$

Exercise 5.1

1. In a journal bearing with a 200 mm length, a 120 mm-diameter shaft rotates at 150 revolutions per minute. If the two surfaces are evenly spaced apart by 0.6 mm and the lubricating oil has a linear velocity distribution with a dynamic viscosity of 0.045 poise, determine the

   (i) Amount of torque needed to overcome the oil's viscous resistance in the journal bearing and
   (ii) Power that the bearing absorbed.

2. With its lower end in a foot-step bearing, a vertical shaft with a diameter of 150 mm rotates at 650 revolutions per minute. If the shaft's end and the bearing's surface are both flat and separated by an oil film that is 0.6 mm thick and has a dynamic viscosity of 1.56 poise, calculate the

   (i) Amount of torque necessary to overcome the oil's viscous resistance in the foot-step bearing and
   (ii) Power necessary to rotate the vertical shaft.

3. The thrust of a shaft rotating at 300 rpm is taken by a collar bearing having internal and external dimensions of 180 mm and 240 mm, respectively. If a lubricating oil film of 0.5 mm thickness and 0.85 poise is maintained between the collar and the bearing, Calculate the

   (i) Torque needed to overcome the collar bearing's oil's viscous resistance and
   (ii) Power loss in overcoming the oil's viscous resistance.

5.5.2   Gears

A *gear* is a machine element that operates in pairs (having geometric compatibility, i.e., the same type and same module) with projections, known as teeth, for maintaining positive engagement between both gears [24]. Gears are rugged, strong and compact tribo-pairs for torque-transmitting. Bending and contact stresses should be less than allowable stresses for gear systems to operate successfully and last for decades when designed and maintained properly. In practice, gear strength deteriorates over time as a result of constant gear material wear. Thus, to maintain a good service life, chemically inert gear material and suitable operating conditions (at an ultra-low wear rate of 10^{-10} mm/min) are needed [24]. The fact that sliding occurs at the surfaces of gear teeth, despite the fact that gears are intended to provide positive drive without slip, is noteworthy. This sliding contributes to one major source of power loss and results in the tribological failure (i.e., wear, scuffing and pitting) of gears [9].

5.6 Summary

A lubricant is any class of substance that can be liquid, solid, or exceptionally gaseous films interposed between interfacing surfaces in relative motion to minimise friction and protect against wear.

The principal lubricant types are liquid-oily lubricant—a film of mostly liquid that comprises organic or inorganic substances such as mineral oils, synthetic esters, silicone fluids and water; and solid lubricant—a film of solid substance that comprises organic or inorganic metallic compounds. The three basic kinds of inorganic compounds that are used as solid lubricants are layer-lattice solids, miscellaneous soft solids and chemical conversion coatings; and gaseous lubricants—a film of gaseous materials composed of substances such as air, industrial gases and liquid–metal vapours.

The predominant commercial fluid lubricant properties include viscosity—a fluid lubricant's property that causes it to resist flowing under load; pour point—the minimum temperature at which a fluid pours or flows when chilled without disturbance at prescribed conditions; flash and fire point—the temperature at which a fluid releases sufficient vapour at its surface to ignite when exposed to an open flame; oiliness—a fluid's tendency to wet and adhere to surfaces; neutralisation number—a measure of the acid or alkaline content of the new fluid or an indicator of the spent fluid's degree of oxidation degradation; and penetration number—a measure of a grease's film characteristics.

The reasons the effectiveness of the lubricating oil is lost (i.e., deteriorates) over time or the causes of oil degradation include viscosity loss, oxidation, thermal breakdown, additive depletion, the formation of condensate and fuel dilution.

An additive is a compound, either organic or inorganic, suspended as solids or dissolved in lubricating oils to impart specific properties to the finished oils. The basic roles of additives are to improve the properties of base oil, suppress the undesirable properties of base oil and impart the base oil with new properties. The more commonly used additives include pour-point depressants, viscosity index improvers, defoamants, oxidation inhibitors, rust and corrosion inhibitors, detergents and dispersants, antiwear additives, and extreme pressure additives, among others.

References

1. Pirro, D. M., Webster, M., & Daschner, E. (2016). *Lubrication fundamentals* (3rd Ed., Revised and Expanded). Routledge, Taylor & Francis Group.
2. Pirro, D. M., & Wessol, A. A. (2001). *Lubrication fundamentals* (2nd Ed., Revised and Expanded). Marcel Dekker, Inc.
3. Harnoy, A. (2003). *Selection and design of rolling bearings.* Marcel Dekker Inc.
4. Patel, S., & Deheri, G. (2019). *Computational modelling of some tribological problems in lubrication.* LAP Lambert Academic Publishing.
5. Abdelbary, A., & Chang, L. (2023). *Principles of engineering tribology: Fundamentals and applications.* Academic Press. https://doi.org/10.1016/B978-0-323-99115-5.01001-X

6. Britannica, T. (Ed.). (2023). Lubrication. *Encyclopedia Britannica*. Retrieved from https://www.britannica.com/technology/lubrication

7. Speight, J. G. (2015). *Handbook of petroleum product analysis* (2nd ed.). Wiley.

8. Meng, T. (Ed.). (2014). *Encyclopedia of lubricants and lubrication*. Springer.

9. Wang, O. J., & Chung, Y. (Eds.). (2013). *Encyclopedia of tribology*. Springer.

10. Meng, T., & Dresel, W. (2007). *Lubricants and lubrication*. Wiley-VCH.

11. Scott, R., Fitch, J., & Leugner, L. (2012). *The practical handbook of machinery lubrication* (4th ed.). Noria Corporation.

12. Bloch, H. P. (2009). *Practical lubrication for industrial facilities*. Fairmont Press Inc.

13. Bhushan, B., Israelachvili, J. N., & Landman, U. (1995). Nanotribology: Friction, wear and lubrication at the atomic scale. *Nature, 374*(6523), 607–616.

14. Ajimotokan, H. A. (2023). *MEE 344: Tribology. Course guide*. Department of Mechanical Engineering, University of Ilorin, Ilorin, Nigeria.

15. Rand, S. J. (Ed.) (2003). *The significance of tests for petroleum products*. ASTM International.

16. Davim, J. P. (Ed.). (2017). *Progress in green tribology: Green and conventional techniques*. De Gruyter Series in Advanced Mechanical Engineering, Vol. 2. De Gruyter. https://doi.org/10.1515/9783110367058

17. Ahmed, N. S., & Nassar, A. M. (2011). Lubricating oil additives. In C. Kuo (Ed.), *Tribology: Lubricants and lubrication*. InTech Europe. Retrieved from http://www.intechopen.com/books/tribology-lubricants-and-lubrication/lubricating-oil-additives

18. Rizvi, S. Q. A. (2009). *A comprehensive review of lubricant chemistry, technology, selection and design*. ASTM International.

19. Leslie, R. R. (2003). *Lubricant additives: Chemistry and applications*. Marcel Dekker Inc.

20. Speight, J., & Exall, D. I. (2014). *Refining used lubricating oils* (1st ed.). CRC Press.

21. Möller, U. J., & Young, D. G. (2003). Hydraulic fluids. *Ullmann's Encyclopedia of Industrial Chemistry*. Wiley Online Library. https://doi.org/10.1002/14356007.a13_165.pub2

22. Margareth, J. S., Peter, R. S., Carlos, R. P. B., & José, R. S. (2010). Lubricant viscosity and viscosity improver additive effects on diesel fuel economy. *Tribology International, 43*, 2298–2302.

23. Kyunghyun, R. (2010). The characteristics of performance and exhaust emissions of a diesel engine using a bio-diesel with antioxidants. *Bioresource Technology, 101*, 578–582.

24. Hirani, H. (2016). *Fundamentals of engineering tribology with applications*. Cambridge University Press.

25. Doll, G. L. (2023). *Rolling bearing tribology: Tribology and failure modes of rolling element bearings*. Elsevier. https://doi.org/10.1016/B978-0-12-822141-9.09993-3

26. Rajput, R. K. (1998). *Fluid mechanics and hydraulic machines*. S. Chand & Company Ltd.

Printed in the United States
by Baker & Taylor Publisher Services